Praise for Joshua Foer's *Moonwalking with Einstein*

"Captivating . . . His narrative is smart and funny and, like the work of Dr. Oliver Sacks, it's informed by a humanism that enables its author to place the mysteries of the brain within a larger philosophical and cultural context."
—Michiko Kakutani, *The New York Times*

"His passionate and deeply engrossing book . . . is a resounding tribute to the muscularity of the mind. . . . In the end, *Moonwalking with Einstein* reminds us that though brain science is a wild frontier and the mechanics of memory little understood, our minds are capable of epic achievements."
—*The Washington Post*

"Joshua Foer's book . . . is both fun and reassuring. All it takes to have a better memory, he contends, are a few tricks and a good erotic imagination."
—Maureen Dowd, *The New York Times*

"Highly entertaining." —Adam Gopnik, *The New Yorker*

"It's delightful to travel with him on this unlikely journey, and his entertaining treatment of memory as both sport and science is spot on. . . . *Moonwalking with Einstein* proves uplifting: It shows that with motivation, focus, and a few clever tricks, our minds can do rather extraordinary things."
—*The Wall Street Journal*

"It's a terrific book: sometimes weird but mostly smart, funny, and ultimately a lovely exploration of the ways that we preserve our lives and our world in the golden amber of human memory." —Deborah Blum, *New Scientist*

"Foer's book is relevant and entertaining as he shows us ways we can unlock our own talent to remember more." —*USA Today*

"A fascinating scientific analysis of mnemonic mysteries. What we remember, [Foer] says, defines who we are." —*Entertainment Weekly*

"Sprightly, entertaining . . . [Foer] has a gift for communicating fairly complex ideas in a manner that is palatable without being patronizing."
—*Financial Times*

"[An] inspired and well-written debut book about not just memorization, but about what it means to be educated and the best way to become so, about expertise in general, and about the not-so-hidden 'secrets' of acquiring skills."
—*The Seattle Times*

"[An] instant bestseller." —*San Francisco Chronicle*

"Funny, curious, erudite, and full of useful details about ancient techniques of training memory." —*The Boston Globe*

"With originality, high energy, and an appealing blend of chutzpah and humility, [Foer] writes of his own adventures and probes the history and literature of memory, the science of how the brain functions, and the connections between memory, identity, and culture. . . . *Moonwalking with Einstein* . . . is engaging and timely." —*The Jewish Week*

"A smart, thoughtful, engaging book." —*The Portland Oregonian*

"Charming . . . The book is part of a grand tradition, the writer as participating athlete, reminiscent of George Plimpton taking up football in *Paper Lion*." —*O, The Oprah Magazine*

"[A] wonderful first book." —*Newcity*

"Fascinating." —*Town & Country*

"For one year, Foer tried to attain total recall, extracting secrets from the top researchers, the real Rain Man, and the world's memory champs. He triumphed, both in his quest and in this lively account, which is, no exaggeration, unforgettable." —*Parade*

"In recounting his year in training for the USA Memory Championship, journalist Foer delivers a rich history of memory." —*Discover Magazine*

"Foer's history of memory is rich with information about the nature of memory and how it makes us who we are." —*Scientific American*

"A brief and pithy recounting of Foer's exploration of the fuzzy borders of his brain—a marveling at how and why it's able to do something quite unexpected. . . . *Moonwalking with Einstein* fits handily inline with the recent tradition of 'big idea' books." —*The Millions*

"An original, entertaining exploration about how and why we remember." —*Kirkus Reviews*

"An engaging, informative, and for the forgetful, encouraging book." —*Booklist*

"Hard to put down . . . The mind is a bigger thing than any of us realize, and Foer reminds us to keep exploring it."

—*Barnes & Noble Review*

"He has thought deeply about memory and his effort yields questions that are well worth reflecting on."

—*The Daily Beast*

"Intriguing . . . Foer does an excellent job of tracing the history of the arts of memory."

—*The Forward*

"The kind of nonfiction work that gets people talking . . . A highly enjoyable read."

—Thirteen.org

"You have to love a writer who employs chick-sexing to help explain human memory. Foer is a charmer, a crackling mind, a fresh wind. He approaches a complex topic with so much humanity, humor, and originality that you don't realize how much you're taking in and understanding. It's kind of miraculous."

—Mary Roach, author of *Packing for Mars*, *Bonk*, *Spook*, and *Stiff*

"*Moonwalking with Einstein* isn't just a splendid overview of an essential aspect of our humanity—our memory; it is also a witty and engaging account of how Foer went from being a guy with an average memory to winning the USA Memory Championship."

—Dan Ariely, professor of behavioral economics at Duke University and author of *The Upside of Irrationality* and *Predictably Irrational*

"In this marvelous book, Joshua Foer invents a new genre of nonfiction. This is a work of science journalism wrapped around an adventure story, a bildungsroman fused to a vivid investigation of human memory. If you want to understand how we remember, and how we can all learn to remember better, then read this book."

—Jonah Lehrer, contributing editor to *Wired* and author of *How We Decide* and *Proust Was a Neuroscientist*

"Joshua Foer proves what few of us are willing to get our heads around: there's more room in our brains than we ever imagined. *Moonwalking with Einstein* isn't a how-to guide to remembering a name or where you put your keys. It's a riveting exploration of humankind's centuries-old obsession with memory, and one man's improbable quest to master his own."

—Stefan Fatsis, author of *A Few Seconds of Panic* and *Word Freak*

ABOUT THE AUTHOR

Joshua Foer has written for *National Geographic*, *Esquire*, *The New York Times*, *The Washington Post*, and *Slate*.

www.joshuafoer.com
jf@joshuafoer.com

MOONWALKING WITH EINSTEIN

The Art and Science of Remembering Everything

JOSHUA FOER

PENGUIN BOOKS

PENGUIN BOOKS
Published by the Penguin Group
Penguin Group (USA) Inc., 375 Hudson Street, New York, New York 10014, U.S.A. * Penguin Group (Canada), 90 Eglinton Avenue East, Suite 700, Toronto, Ontario, Canada M4P 2Y3 (a division of Pearson Penguin Canada Inc.) * Penguin Books Ltd, 80 Strand, London WC2R 0RL, England * Penguin Ireland, 25 St. Stephen's Green, Dublin 2, Ireland (a division of Penguin Books Ltd) * Penguin Books Australia Ltd, 250 Camberwell Road, Camberwell, Victoria 3124, Australia (a division of Pearson Australia Group Pty Ltd) * Penguin Books India Pvt Ltd, 11 Community Centre, Panchsheel Park, New Delhi – 110 017, India * Penguin Group (NZ), 67 Apollo Drive, Rosedale, North Shore 0632, New Zealand (a division of Pearson New Zealand Ltd) * Penguin Books (South Africa) (Pty) Ltd, 24 Sturdee Avenue, Rosebank, Johannesburg 2196, South Africa

Penguin Books Ltd, Registered Offices:
80 Strand, London WC2R 0RL, England

First published in the United States of America by The Penguin Press,
a member of Penguin Group (USA) Inc. 2011
Published in Penguin Books 2012

31st Printing

THE LIBRARY OF CONGRESS HAS CATALOGED THE HARDCOVER EDITION AS FOLLOWS:
Foer, Joshua.
Moonwalking with Einstein : the art and science
of remembering everything / Joshua Foer.
p. cm.
Includes bibliographical references and index.
ISBN 978-1-59420-229-2 (hc.)
ISBN 978-0-14-312053-7 (pbk.)
1. Mnemonics. 2. Memory. I. Title.
BF385.F64 2011
153.1'4—dc22 2010030265

Printed in the United States of America
DESIGNED BY MEIGHAN CAVANAUGH

For Dinah: Everything.

CONTENTS

MOONWALKING

WITH

EINSTEIN

There were no other survivors.

Family members arriving at the scene of the fifth-century-B.C. banquet hall catastrophe pawed at the debris for signs of their loved ones—rings, sandals, anything that would allow them to identify their kin for proper burial.

Minutes earlier, the Greek poet Simonides of Ceos had stood to deliver an ode in celebration of Scopas, a Thessalian nobleman. As Simonides sat down, a messenger tapped him on the shoulder. Two young men on horseback were waiting outside, anxious to tell him something. He stood up again and walked out the door. At the very moment he crossed the threshold, the roof of the banquet hall collapsed in a thundering plume of marble shards and dust.

He stood now before a landscape of rubble and entombed bodies. The air, which had been filled with boisterous laughter moments before, was smoky and silent. Teams of rescuers set to work frantically digging through the collapsed building. The corpses they pulled out of the wreckage were mangled beyond recognition. No one could even say for sure who had been inside. One tragedy compounded another.

Then something remarkable happened that would change forever how people thought about their memories. Simonides sealed his senses to the chaos around him and reversed time in his mind. The piles of marble returned to pillars and the scattered frieze fragments reassembled in the air above. The stoneware scattered in the debris re-formed into bowls. The splinters of wood poking above the ruins once again became a table. Simonides caught a glimpse of each of the banquet guests at his seat, carrying on oblivious to the impending catastrophe. He saw Scopas laughing at the head of the table, a fellow poet sitting across from him sponging up the remnants of his meal with a piece of bread, a nobleman smirking. He turned to the window and saw the messengers approaching, as if with some important news.

Simonides opened his eyes. He took each of the hysterical relatives by the hand and, carefully stepping over the debris, guided them, one by one, to the spots in the rubble where their loved ones had been sitting.

At that moment, according to legend, the art of memory was born.

ONE

THE SMARTEST MAN IS HARD TO FIND

Dom DeLuise, celebrity fat man (and five of clubs), has been implicated in the following unseemly acts in my mind's eye: He has hocked a fat globule of spittle (nine of clubs) on Albert Einstein's thick white mane (three of diamonds) and delivered a devastating karate kick (five of spades) to the groin of Pope Benedict XVI (six of diamonds). Michael Jackson (king of hearts) has engaged in behavior bizarre even for him. He has defecated (two of clubs) on a salmon burger (king of clubs) and captured his flatulence (queen of clubs) in a balloon (six of spades). Rhea Perlman, diminutive *Cheers* bartendress (and queen of spades), has been caught cavorting with the seven-foot-seven Sudanese basketball star Manute Bol (seven of clubs) in a highly explicit (and in this case, anatomically improbable) two-digit act of congress (three of clubs).

This tawdry tableau, which I'm not proud to commit to the page, goes a long way toward explaining the unlikely spot I find myself in at the moment. Sitting to my left is Ram Kolli, an unshaven twenty-five-year-old business consultant from Richmond, Virginia, who is also the defending United States memory champion. To my right is the muzzle of a television camera from a national cable network. Spread out behind me, where I can't see them and they can't disturb me, are about a hundred spectators and a pair of TV commentators offering play-by-play analysis. One is a blow-dried veteran boxing announcer named Kenny Rice, whose gravelly, bedtime voice can't conceal the fact that he seems bewildered by this jamboree of nerds. The other is the Pelé of USA memory sport, a bearded forty-three-year-old chemical engineer and four-time national champion from Fayetteville, North Carolina, named Scott Hagwood. In the corner of the room sits the object of my affection: a kitschy two-tiered trophy consisting of a silver hand with gold nail polish brandishing a royal flush, and, in a patriotic flourish, three bald eagles perched just below. It's nearly as tall as my two-year-old niece (and lighter than most of her stuffed animals).

The audience has been asked not to take any flash photographs and to maintain total silence. Not that Ram or I could possibly hear them. Both of us are wearing earplugs. I've also got on a pair of industrial-strength earmuffs that look like they belong on an aircraft carrier deckhand (because in the heat of a memory competition, there is no such thing as deaf enough). My eyes are closed. On a table in front of me, lying facedown between my hands, are two shuffled decks of playing cards. In a moment, the chief arbiter will click a stopwatch and I will have five minutes to memorize the order of both decks.

The unlikely story of how I ended up in the finals of the USA Memory Championship, stock-still and sweating profusely, begins a year earlier on a snowy highway in central Pennsylvania. I had been driving from my home in Washington, D.C., to the Lehigh Valley to do an interview for *Discover* magazine with a theoretical physicist at Kutztown University, who had invented a vacuum chamber device that was supposed to pop the world's largest popcorn. My route took me through York, Pennsylvania, home of the Weightlifting Hall of Fame and Museum. I thought that sounded like something I didn't want to die without having seen. And I had an hour to kill.

As it turned out, the Hall of Fame was little more than a sterile collection of old photographs and memorabilia displayed on the ground floor of the corporate offices of the nation's largest barbell manufacturer. Museologically, it was crap. But it's where I first saw a black-and-white photo of Joe "The Mighty Atom" Greenstein, a hulking five-foot-four Jewish-American strongman who had earned his nickname in the 1920s by performing such inspiring feats as biting quarters in half and lying on a bed of nails while a fourteen-man Dixieland band played on his chest. He once changed all four tires on a car without any tools. A caption next to the photo billed Greenstein as "the strongest man in the world."

Staring at that photo, I thought it would be pretty interesting if the world's strongest person ever got to meet the world's smartest person. The Mighty Atom and Einstein, arms wrapped around each other: an epic juxtaposition of muscle and mind. A neat photo to hang above my desk, at least. I wondered if it had ever been taken. When I got home, I did a little Googling. The world's strongest person was pretty easy to find: His name was Mariusz Pudzianowski. He lived in Biała Rawska, Poland, and could deadlift 924 pounds (about thirty of my nieces).

The world's smartest person, on the other hand, was not so easily identified. I typed in "highest IQ," "intelligence champion," "smartest in the world." I learned that there was someone in New York City with an IQ of 228, and a chess player in Hungary who once played fifty-two simultaneous blindfolded games. There was an Indian woman who could calculate the twenty-third root of a two-hundred-digit number in her head in fifty seconds, and someone else who could solve a four-dimensional Rubik's cube, whatever that is. And of course there were plenty of more obvious Stephen Hawking types of candidates. Brains are notoriously trickier to quantify than brawn.

In the course of my Googling, though, I did discover one intriguing candidate who was, if not the smartest person in the world, at least some kind of freakish genius. His name was Ben Pridmore, and he could memorize the precise order of 1,528 random digits in an hour and—to impress those of us with a more humanist bent—any poem handed to him. He was the reigning world memory champion.

Over the next few days, my brain kept returning to Ben Pridmore's. My own memory was average at best. Among the things I regularly forgot: where I put my car keys (where I put my car, for that matter); the food in the oven; that it's "its" and not "it's"; my girlfriend's birthday, our anniversary, Valentine's Day; the clearance of the doorway to my parents' cellar (ouch); my friends' phone numbers; why I just opened the fridge; to plug in my cell phone; the name of President Bush's chief of staff; the order of the New Jersey Turnpike rest stops; which year the Redskins last won the Super Bowl; to put the toilet seat down.

Ben Pridmore, on the other hand, could memorize the order of a shuffled deck of playing cards in thirty-two seconds. In five minutes he could permanently commit to memory what happened on ninety-six different historical dates. The man knew fifty thousand digits of pi. What was not to envy? I had once read that the average person squanders about forty days a year compensating for things he or she has forgotten. Putting aside for a moment the fact that he was temporarily unemployed, how much more productive must Ben Pridmore be?

Every day there seems to be more to remember: more names, more passwords, more appointments. With a memory like Ben Pridmore's, I imagined, life would be qualitatively different—and better. Our culture constantly inundates us with new information, and yet our brains capture so little of it. Most just goes in one ear and out the other. If the point of reading were simply to retain knowledge, it would probably be the single least efficient activity I engage in. I can spend a half dozen hours reading a book and then have only a foggy notion of what it was about. All those facts and anecdotes, even the stuff interesting enough to be worth underlining, have a habit of briefly making an impression on me and then disappearing into who knows where. There are books on my shelf that I can't even remember whether I've read or not.

What would it mean to have all that otherwise-lost knowledge at my fingertips? I couldn't help but think that it would make me more persuasive, more confident, and, in some fundamental sense, smarter. Certainly I'd be a better journalist, friend, and boyfriend. But more than that, I imagined that having a memory like Ben Pridmore's would make me an altogether more attentive, perhaps even wiser, person. To the extent that experience is the sum of our memories and wisdom the sum of experience, having a better memory would mean knowing not only more about the world, but also more about myself. Surely some of the forgetting that seems to plague us is healthy and necessary. If I didn't forget so many of the dumb things I've done, I'd probably be unbearably neurotic. But how many worthwhile ideas have gone unthought and connections unmade because of my memory's shortcomings?

I kept returning to something Ben Pridmore said in a newspaper interview, which made me ponder just how different his memory and my own might really be. "It's all about technique and understanding how the memory works," he told the reporter. "Anyone could do it, really."

. . .

A couple weeks after my trip to the Weightlifting Hall of Fame, I stood in the back of an auditorium on the nineteenth floor of the Con Edison headquarters near Union Square in Manhattan, an observer at the 2005 USA Memory Championship. Spurred by my fascination with Ben Pridmore, I was there to write a short piece for *Slate* magazine about what I imagined would be the Super Bowl of savants.

The scene I stumbled on, however, was something less than a clash of titans: a bunch of guys (and a few ladies), widely varying in both age and hygienic upkeep, poring over pages of random numbers and long lists of words. They referred to themselves as "mental athletes," or just MAs for short.

There were five events. First the contestants had to learn by heart a fifty-line unpublished poem called "The Tapestry of Me." Then they were provided with ninety-nine photographic head shots accompanied by first and last names and given fifteen minutes to memorize as many of them as possible. Then they had another fifteen minutes to memorize a list of three hundred random words, five minutes to memorize a page of a thousand random digits (twenty-five lines of numbers, forty numbers to a line), and another five minutes to learn the order of a shuffled deck of playing cards. Among the competitors were two of the world's thirty-six grand masters of memory, a rank attained by memorizing a sequence of a thousand random digits in under an hour, the precise order of ten shuffled decks of playing cards in the same amount of time, and the order of one shuffled deck in less than two minutes.

Though on the face of it these feats might seem like little more than geeky party tricks—essentially useless, and perhaps even vaguely pathetic—what I discovered as I talked to the competitors was something far more serious, a story that made me reconsider the limits of my own mind and the very essence of my education.

I asked Ed Cooke, a young grand master from England who had come to the USA event as spring training for that summer's World Championship (since he was a non-American, his scores couldn't be counted in the USA contest), when he first realized he was a savant.

"Oh, I'm not a savant," he said, chuckling.

"Photographic memory?" I asked.

He chuckled again. "Photographic memory is a detestable myth," he said. "Doesn't exist. In fact, my memory is quite average. All of us here have average memories."

That seemed hard to square with the fact that I'd just watched him recite back 252 random digits as effortlessly as if they'd been his own telephone number.

"What you have to understand is that even average memories are remarkably powerful if used properly," he said. Ed had a blocky face and a shoulder-length mop of curly brown hair, and could be counted among the competitors who were least concerned with habits of personal grooming. He was wearing a suit with a loosened tie and, incongruously, a pair of flip-flops emblazoned with the Union Jack. He was twenty-four years old but carried his body like someone three times that age. He hobbled about with a cane at his side—"a winning prop," he called it—which was necessitated by a recent painful relapse of chronic juvenile arthritis. He and all the other mental athletes I met kept insisting, as Ben Pridmore had in his interview, that anyone could do what they do. It was simply a matter of learning to "think in more memorable ways" using the "extraordinarily simple" 2,500-year-old mnemonic technique known as the "memory palace" that Simonides of Ceos had supposedly invented in the rubble of the great banquet hall collapse.

The techniques of the memory palace—also known as the journey method or the method of loci, and more broadly as the *ars memorativa,* or "art of memory"—were refined and codified in an extensive set of rules and instruction manuals by Romans like Cicero and Quintilian, and flowered in the Middle Ages as a way for the pious to memorize everything from sermons and prayers to the punishments awaiting the wicked in hell. These were the same tricks that Roman senators had used to memorize their speeches, that the Athenian statesman Themistocles had supposedly used to memorize the names of twenty thousand Athenians, and that medieval scholars had used to memorize entire books.

Ed explained to me that the competitors saw themselves as "participants in an amateur research program" whose aim was to rescue a long-lost tradition of memory training that had disappeared centuries ago. Once upon a time, Ed insisted, remembering was everything. A trained memory was not just a handy tool, but a fundamental facet of any worldly mind. What's more, memory training was considered a form of character building, a way of developing the cardinal virtue of prudence and, by extension, ethics. Only through memorizing, the thinking went, could ideas truly be incorporated into one's psyche and their values absorbed. The techniques existed not just to memorize useless information like decks of playing cards, but also to etch into the brain foundational texts and ideas.

But then, in the fifteenth century, Gutenberg came along and turned books into mass-produced commodities, and eventually it was no longer all that important to remember what the printed page could remember for you. Memory techniques that had once been a staple of classical and medieval culture got wrapped up with the occult and esoteric Hermetic traditions of the Renaissance, and by the nineteenth century they had been relegated to carnival sideshows and tacky self-help books—only to be resurrected in the last decades of the twentieth century for this bizarre and singular competition.

The leader of this renaissance in memory training is a slick sixty-seven-year-old British educator and self-styled guru named Tony

Buzan, who claims to have the highest "creativity quotient" in the world. When I met him, in the cafeteria of the Con Edison building, he was wearing a navy suit with five enormous gold-rimmed buttons and a collarless shirt, with another large button at his throat that gave him the air of an Eastern priest. A neuron-shaped pin adorned his lapel. His watch face bore a reproduction of Dali's *Persistence of Memory* (the one with the dripping watch face). He referred to the competitors as "warriors of the mind."

Buzan's grizzled face looked a decade older than his sixty-seven years, but the rest of him was as trim as a thirty-year-old. He rows between six and ten kilometers every morning on the river Thames, he told me, and he makes a point of eating lots of "brain-healthy" vegetables and fish. "Junk food in: junk brain. Healthy food in: healthy brain," he said.

When he walked, Buzan seemed to glide across the floor like an air hockey puck (the result, he later told me, of forty years' training in the Alexander Technique). When he spoke, he gesticulated with a polished, staccato precision that could only have been honed in front of a mirror. Often, he punctuated a key point with an explosion of fingers from his closed fist.

Buzan founded the World Memory Championship in 1991 and has since established national championships in more than a dozen countries, from China to South Africa to Mexico. He says he has been working with a missionary's zeal since the 1970s to get these memory techniques implemented in schools around the world. He calls it a "global education revolution focusing on learning how to learn." And he's been minting himself a serious fortune in the process. (According to press reports, Michael Jackson ran up a $343,000 bill for Buzan's mind-boosting services shortly before his death.)

Buzan believes schools go about teaching all wrong. They pour vast amounts of information into students' heads, but don't teach them how to retain it. Memorizing has gotten a bad rap as a mindless way of holding onto facts just long enough to pass the next exam. But it's not memorization that's evil, he says; it's the tradition of boring rote learning that he believes has corrupted Western education. "What we have been doing over the last century is defining memory incorrectly, understanding it incompletely, applying it inappropriately, and condemning it because it doesn't work and isn't enjoyable," Buzan argues. If rote memorization is a way of scratching impressions onto our brains through the brute force of repetition—the old "drill and kill" method—then the art of memory is a more elegant way of remembering through technique. It is faster, less painful, and produces longer-lasting memories, Buzan told me.

"The brain is like a muscle," he said, and memory training is a form of mental workout. Over time, like any form of exercise, it'll make the brain fitter, quicker, and more nimble. It's an idea that dates back to the very origins of memory training. Roman orators argued that the art of memory—the proper retention and ordering of knowledge—was a vital instrument for the invention of new ideas. Today, the "mental workout" has gained great currency in the popular imagination. Brain gyms and memory boot camps are a growing fad, and brain training software was a $265 million industry in 2008, no doubt in part because of research that shows that older people who keep their minds active with crossword puzzles and chess can stave off Alzheimer's and progressive dementia, but mostly because of the Baby Boomer generation's intense insecurity about losing their marbles. But while there is much solid science to back up the dementia-defying benefits of an active brain, Buzan's most hyperbolic claims about the collateral effects of "brain exercise" should inspire a measured dose (at least) of skepticism. Nevertheless, it was hard to argue with the results. I'd just watched a forty-seven-year-old competitor recite, in order, a list of a hundred random words he'd learned a few minutes earlier.

Buzan was eager to sell me on the idea that his own memory has been improving year after year, even as he ages. "People assume that memory decline is a function of being human, and therefore natural," he said. "But that is a logical error, because normal is not necessarily natural. The reason for the monitored decline in human memory performance is because we actually do anti-Olympic training. What we do to the brain is the equivalent of sitting someone down to train for the Olympics and making sure he drinks ten cans of beer a day, smokes fifty cigarettes, drives to work, and maybe does some exercise once a month that's violent and damaging, and spends the rest of the time watching television. And then we wonder why that person doesn't do well in the Olympics. That's what we've been doing with memory."

I pestered Buzan about how hard it would be to learn these techniques. How did the competitors train? How quickly did their memories improve? Did they use these techniques in everyday life? If they were really as simple and effective as he was claiming, how come I'd never heard of them before? Why weren't we all using them?

"You know," he replied, "instead of asking me all these questions, you should just try it for yourself."

"What would it take, theoretically, for someone like me to train for the USA Memory Championship?" I asked him.

"If you want to make it into the top three of the U.S. championship, it'd be a good idea to spend an hour a day, six days a week. If you spent that much time, you'd do very well. If you wanted to enter the world championship, you'd need to spend three to four hours a day for the final six months leading up to the championship. It gets heavy."

Later that morning, while the competitors were trying to memorize "The Tapestry of Me," Buzan took me aside and put his hand on my shoulder.

"Remember our little talk? Think about it. That could be you up there on the stage, the next USA memory champion."

. . .

During a break between the poem memorization and the names-and-faces event, I headed for the sidewalk outside the Con Ed building to escape the locker-room humidity. There I ran into the mop-haired, cane-toting English mnemonist Ed Cooke and his lanky sidekick, the Austrian grand master Lukas Amsüss, rolling their own cigarettes.

Ed had graduated from Oxford the previous spring with a first-class degree in psychology and philosophy and told me that he was simultaneously toying with writing a book titled *The Art of Introspection* and pursuing his cognitive science PhD at the University of Paris, where he was doing outré research with the aim of "making people feel like their body has shrunk to a tenth of its normal size." He was also working on inventing a new color—"not just a new color, but a whole new way of seeing color."

Lukas, a University of Vienna law student who advertised himself as the author of a short pamphlet titled "How to Be Three Times Clev-erer Than Your IQ," was leaning against the building, trying to justify to Ed his miserable showing in the random words event: "I've never heard even of these English words 'yawn,' 'ulcer,' and 'aisle' before," he insisted in a stiff Austrian accent, "How can I be expected to memorize them?"

At the time, Ed and Lukas were respectively the eleventh- and ninth-best memorizers in the world, the only grand masters at the event, and the only competitors who had shown up in suit and tie. They were eager to share with me (or anyone) their plan to cash in on their mnemonic fame by building a "memory gymnasium" called the Oxford Mind Academy. Their idea was that subscribers—mostly business executives, they hoped—would pay to have personal mental workout trainers. Once the world learned the benefits of training one's memory, they imagined that cash would fall from the sky. "Ultimately," Ed told me, "we are looking to rehabilitate Western education."

"Which we consider to be degenerate," Lukas added.

Ed explained to me that he saw his participation in memory competitions as part of his attempt to unravel the secrets of human memory. "I figure that there are two ways of figuring out how the brain works," he said. "The first is the way that empirical psychology does it, which is that you look from the outside and take a load of measurements on a load of different people. The other way follows from the logic that a system's optimal performance can tell you something about its design. Perhaps the best way to understand human memory is to try very hard to optimize it—ideally with a load of bright people in conditions where they get rigorous and objective feedback. That's what the memory circuit is."

The contest itself unfolded with all the excitement of, say, the SAT. The contestants sat quietly at tables staring at sheets of paper, then scribbled answers that they handed off to judges. After each event, scores were quickly calculated and displayed on a screen at the front of the room. But much to the dismay of a journalist trying to write about a national memory championship, the "sport" had none of the public agony of a basketball game, or even a spelling bee. Sometimes it was difficult to tell whether competitors were deep in thought or sleep. There may have been a lot of dramatic temple massaging and nervous foot tapping and the occasional empty stare of defeat, but mostly the drama was inside the competitors' heads, inaccessible to spectators.

A troubling thought percolated to the front of my brain as I stood in the back of the Con Edison auditorium watching these supposedly normal human beings perform their almost unfathomable mental acrobatics: I didn't have a clue how my own memory worked. Was there even such a place as the front of my brain? A slow wave of questions swept over me—things I'd never bothered to wonder about, but which all of a sudden seemed profoundly pressing. What exactly *is* a memory? How is one created? And how does it get stored? I'd spent the first two and a half decades of my life with a memory that operated so seamlessly that I'd never had cause to stop and inquire about its mechanics. And yet, now that I was stopping to think about it, I realized that it actually didn't work all that seamlessly. It completely failed in certain areas, and worked far too well in others. And it had so many inexplicable quirks. That very morning my brain had been held hostage by an unbearable Britney Spears song, forcing me to spend the better part of a subway ride humming Hanukkah jingles in an attempt to dislodge it. What was that about? A few days earlier, I had been trying to tell a friend about an author I admired, only to find that I remembered the first letter of his last name, and nothing else. How come that happened? And why didn't I have a single memory before the age of three? For that matter, why couldn't I remember what I had for breakfast just the day before, even though I remembered exactly what I was having for breakfast—Corn Pops, coffee, and a banana—four years earlier when I was told that a plane had just crashed into one of the twin towers? And why am I always forgetting why I opened the refrigerator door?

I came away from the USA Memory Championship eager to find out how Ed and Lukas did it. Were these just extraordinary individuals, prodigies from the long tail of humanity's bell curve, or was there something we could all learn from their talents? I was skeptical about them for the same reason I was skeptical about Tony Buzan. Any self-appointed guru who has earned himself a king's ransom in the modern "self-help" racket is bound to perk up a journalist's bullshit detector, and Buzan had set off every alarm bell I've got. I didn't yet know enough to know whether he was selling hype or science, but his over-the-top packaging—"a global education revolution"!—certainly smacked of the former.

Was it really true that anyone could learn to quickly memorize huge quantities of information? *Anyone?* I was willing to believe Buzan when he said there were techniques that one could learn to marginally improve one's memory around the edges, but I didn't fully believe him (or Ed) when he said that any schmo off the street could learn to memorize entire decks of playing cards or thousands of binary digits. The alternate explanation just seemed a whole lot more plausible: that Ed and his colleagues had some freakish innate talent that was the mental equivalent of André the Giant's height or Usain Bolt's legs.

Indeed, much of what's been written about memory improvement by self-help gurus is tainted by hucksterism. When I checked out the self-help aisle at my local Barnes & Noble, I found stacks of books making fevered claims that they could teach me how to "never forget a telephone number or date" or "develop instant recall." One book even pronounced that it could show me how to use the "other 90 percent" of my brain, which is one of those pseudoscientific clichés that makes about as much sense as saying I could be taught to use the other 90 percent of my hand.

But memory improvement has also long been investigated by people whose relationships to the subject are less obviously profitable and whose claims are inspected by peer review. Academic psychologists have been interested in expanding our native memory capacities ever since Hermann Ebbinghaus first brought the study of memory into the laboratory in the 1870s.

This book is about the year I spent trying to train my memory, and also trying to understand it—its inner workings, its natural deficiencies, its hidden potential. It's about how I learned firsthand that our memories are indeed improvable, within limits, and that the skills of Ed and Lukas can indeed be tapped by all of us. It's also about the scientific study of expertise, and how researchers who study memory champions have discovered general principles of skill acquisition—secrets to improving at just about anything—from how mental athletes train their brains.

Though this is not meant to be a self-help book, I hope you'll come away with a sense of how one goes about training one's memory, and how memory techniques can be used in everyday life.

Those techniques have a surprisingly rich and important legacy. The role that they have played in the development of Western culture is one of the great themes in intellectual history whose story is not widely known outside of the rarefied academic corners in which it is studied. Mnemonic systems like Simonides' memory palace profoundly shaped the way people approached the world from the time of antiquity through the Middle Ages and the Renaissance. And then they all but disappeared.

Physiologically, we are virtually identical to our ancestors who painted images of bison on the walls of the Lascaux cave in France, among the earliest cultural artifacts to have survived to the present day. Our brains are no larger or more sophisticated than theirs. If one of their babies were to be dropped into the arms of an adoptive parent in twenty-first-century New York, the child would likely grow up indistinguishable from his or her peers.

All that differentiates *us* from *them* is our memories. Not the memories that reside in our own brains, for the child born today enters the world just as much a blank slate as the child born thirty thousand years ago, but rather the memories that are stored outside ourselves—in books, photographs, museums, and these days in digital media. Once upon a time, memory was at the root of all culture, but over the last thirty millennia since humans began painting their memories on cave walls, we've gradually supplanted our own natural memory with a vast superstructure of external memory aids—a process that has sped up exponentially in recent years. Imagine waking up tomorrow and discovering that all the world's ink had become invisible and all our bytes had disappeared. Our world would immediately crumble. Literature, music, law, politics, science, math: Our culture is an edifice built of externalized memories.

If memory is our means of preserving that which we consider most valuable, it is also painfully linked to our own transience. When we die, our memories die with us. In a sense, the elaborate system of externalized memory we've created is a way of fending off mortality. It allows ideas to be efficiently passed across time and space, and for one idea to build on another to a degree not possible when a thought has to be passed from brain to brain in order to be sustained.

The externalization of memory not only changed how people think; it also led to a profound shift in the very notion of what it means to be intelligent. Internal memory became devalued. Erudition evolved from possessing information internally to knowing how and where to find it in the labyrinthine world of external memory. It's a telling statement that pretty much the only place where you'll find people still training their memories is at the World Memory Championship and the dozen national memory contests held around the globe. What was once a cornerstone of Western culture is now at best a curiosity. But as our culture has transformed from one that was fundamentally based on internal memories to one that is fundamentally based on memories stored outside the brain, what are the implications for ourselves and for our society? What we've gained is indisputable. But what have we traded away? What does it mean that we've lost our memory?

TWO

THE MAN WHO REMEMBERED TOO MUCH

In May 1928, the young journalist S walked into the office of the Russian neuropsychologist A. R. Luria and politely asked to have his memory tested. He had been sent by his boss, the editor of the newspaper where he worked. Each morning, at the daily editorial meeting, his boss would dole out the day's assignments to the roomful of reporters in a rapid stream of facts, contacts, and addresses that they would need to file their stories. All the reporters took copious notes, except one. S simply watched and listened.

One morning, fed up at the reporter's apparent inattentiveness, the editor took S aside to lecture him about the need to take his job seriously. Did he think all that information was being read off each morning just because the editor liked to hear his own voice? Did he think he could report his stories without contacts? That he could simply reach out to people telepathically, without knowing their addresses? If he hoped to have any future in the world of newspaper journalism, he'd have to begin paying attention and jotting notes, the editor told him.

S stared at the editor blankly through his scolding and waited for him to finish. Then he calmly repeated back every detail of the morning meeting, word for word. The editor was floored. He didn't know what to say. But S would later claim that he, S, felt the bigger shock. Until that moment, he said, he'd always assumed that it was perfectly normal for a person to remember everything.

Upon arriving at Luria's office, S remained skeptical about his own uniqueness. "He wasn't aware of any peculiarities in himself and couldn't conceive of the idea that his memory differed from other people's," recalled the psychologist, who gave him a series of tests to evaluate his powers of recall. Luria started by asking S to memorize a list of numbers, and listened in amazement as his shy subject recited back seventy digits, first forward and then backward. "It was of no consequence to him whether the series I gave him contained meaningful words or nonsense syllables, numbers or sounds; whether they were presented orally or in writing," said Luria. "All he required was that there be a three-to-four-second pause between each element in the series, and he had no difficulty reproducing whatever I gave him." Luria gave S test after test, and kept getting the same result: The man was unstumpable. "As the experimenter, I soon found myself in a state verging on utter confusion," Luria recalled. "I simply had to admit that . . . I had been unable to perform what one would think was the simplest task a psychologist can do: measure the capacity of an individual's memory."

Luria would go on to study S for the next thirty years, and would eventually write a book about him, *The Mind of a Mnemonist: A Little Book About a Vast Memory*, that has become one of the most enduring classics in the literature of abnormal psychology. S could memorize complex mathematical formulas without knowing any math, Italian poetry without speaking Italian, and even phrases of gobbledygook. But even more remarkable than the breadth of material he could commit to memory was the fact that his memories seemed never to degrade.

For normal humans, memories gradually decay with time along what's known as the "curve of forgetting." From the moment you grasp a new piece of information, your memory's hold on it begins to slowly loosen, until finally it lets go altogether. In the last decades of the nineteenth century, the German psychologist Hermann Ebbinghaus set out to quantify this inexorable process of forgetting. In order to understand how our memories fade over time, he spent years memorizing 2,300 three-letter nonsense syllables like GUF, LER, and NOK. At set periods, he would test himself to see how many of the syllables he'd forgotten and how many he'd managed to retain. When he graphed the results, he got a curve that looked like this:

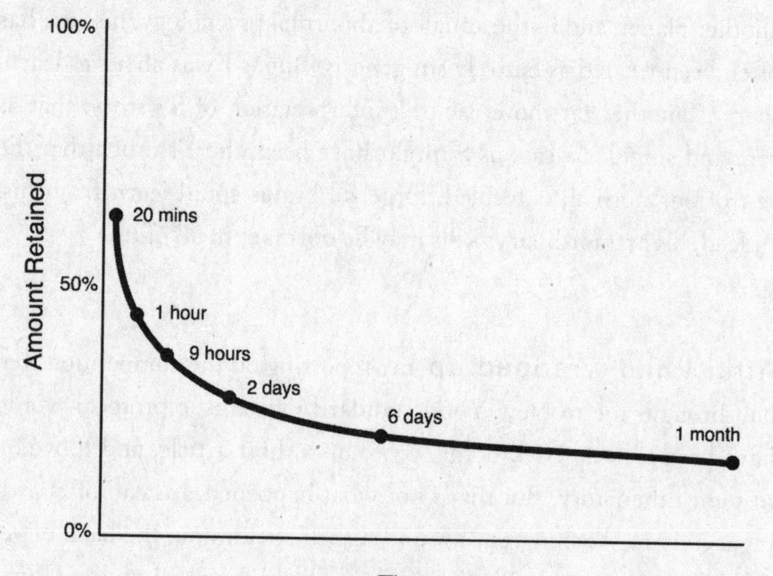

No matter how many times he performed the experiment on himself, the results were always roughly the same: In the first hour after learning a set of nonsense syllables, more than half of them would be forgotten. After the first day, another 10 percent would disappear. After a month, another 14 percent. After that, the memories that were left had more or less stabilized—they had become consolidated in long-term memory—and the pace of forgetting slowed to a gentle creep.

S's memories seemed not to follow the curve of forgetting. No matter how much he'd been asked to remember, or how long ago it had been—as many as sixteen years in some cases—he was always able to recite back the material with the same exactitude as if he'd just learned it. "He would sit with his eyes closed, pause, then comment: 'Yes, yes . . . this was the series you gave me once when we were in your apartment . . . you were sitting at the table . . . you were wearing a grey suit . . .' And with that he would reel off the series precisely as I had given it to him at the earlier session," wrote Luria.

In Luria's lyrical account, S seems at times like a visitor from another planet, and in the annals of abnormal psychology, his case has often been treated as entirely sui generis. But as I was about to learn, there is another far more exciting interpretation of S's story: that as rare and singular a case as S might have been, there's much that the rest of our normal, enfeebled, forgetful brains could learn from his. Indeed, his extraordinary skills may lie dormant in all of us.

After I had wrapped up my reporting on the competition that had brought me to New York, standard journalistic protocol would have been to head back home, write up a short article, and move on to some other story. But that's not what happened. Instead of boarding a train to Washington, I found myself standing in the back of yet another auditorium—this time, at a public high school on the Upper

East Side of Manhattan, where Ed Cooke was supposed to be teaching a roomful of sixteen-year-olds how to use memory techniques to ace their exams. I had canceled my plans for the day and tagged along because he'd promised me that if I hung around with him long enough he would explain to me, in detail, how he and Lukas had taught themselves to remember like S. But before delving into any such esoteric secrets, there was some basic groundwork to be laid. Ed wanted to show me and the students that our memories were already extraordinary—at least when it came to learning certain kinds of information. To do that, he had brought along a version of a memory test known as the two-alternative picture recognition exam.

After introducing himself to the students with some self-deprecating humor—"I'm from England, where we prefer to spend our time memorizing, rather than developing full social lives"—he demonstrated his mnemonic bona fides by learning a seventy-digit number in just over a minute (three times faster than it took S to perform the same feat), and then proceeded straight into a test of the students' memories, and mine.

"I'm going to show you guys a bunch of pictures, and I'm going to show them to you really, really fast," he announced, trying to lift his voice above the clamoring teenagers. "I want you to try to remember as many of them as you can." He pressed a button on a remote control, and the overhead lights dimmed. A series of slides began to blink across a projection screen at the front of the room, each lingering for less than half a second. There was a slide of Muhammad Ali standing triumphantly over Sonny Liston. Then a slide of barbells. Then Neil Armstrong's footprint on the moon. Then the cover of Friedrich Nietzsche's *On the Genealogy of Morals*. And a red rose.

There were thirty such pictures, each appearing and disappearing so quickly that it was hard to imagine we'd ever be able to recall any of them, much less all of them. But I tried my hardest to capture some detail from each, and to make a quick mental note of what I was looking at. After the last slide, a picture of a goat, the wall went blank and the lights came back on.

"Now, do you think you'll be able to remember all those pictures?" Ed asked us.

A girl sitting just in front of me shouted, "Not a chance!" provoking giggles from several of her colleagues.

"That's the spirit!" Ed yelled back, and then looked down at his watch to note the time. Of course, the point of the exercise—why else would he have given it?—was that we *would* be able to remember all those pictures. Like the girl in front of me, I found it hard to believe.

After giving us thirty minutes for the curve of forgetting to work its inevitable erasures on the images we'd glanced at so quickly, Ed put up a new set of slides. This time, there were two pictures on the screen. One of them we'd seen before, and one of them we hadn't: Muhammad Ali on the left and a fizzling Alka-Seltzer tablet on the right.

He asked us all to point to the picture we recognized. Easy enough. We all knew we'd seen Muhammad Ali, but not the Alka-Seltzer tablet. "Isn't it striking how easily you remember that?" said Ed, before clicking through to another slide: a deer on the left and the Nietzsche book on the right.

We all knew that one, too. In fact, he went through thirty slides, and everyone in the room recognized every single one of the photos we'd seen before. "Now here's the fascinating thing," said Ed, pacing professorially at the front of the linoleum-tiled auditorium. "We could have done this with ten thousand slides, and you would have performed almost equally well. Your memory for images is *that* good." He was referring to a frequently cited set of experiments carried out in the 1970s using the exact same picture recognition test that we'd just taken, only instead of thirty images, the researchers asked their subjects to remember ten thousand. (It took five days to perform the test.) That's a lot of pictures for a mind to keep track of, especially since the subjects were only able to look at each image once. Even so, the scientists found that people were able to recognize more than 80 percent of what they'd seen. In a more recent study, the same test was performed with 2,500 images, but instead of asking people to choose between an image of Muhammad Ali and an Alka-Seltzer tablet (an easy choice, no matter how effervescent Cassius Clay might have been), they had to choose between alternative images that were almost identical: a stack of five dollar bills versus a stack of one dollar bills, a green train car versus a red train car, a bell with a narrow handle versus a bell with a wide handle. Even when the images differed only in a tiny detail, people still remembered 90 percent of them correctly.

I found those numbers astonishing, but I realized they were merely quantifying something that I instinctively knew: that our memories do a pretty darn good job. For all of our griping over the everyday failings of our memories—the misplaced keys, the forgotten name, the factoid stuck on the tip of the tongue—their biggest failing may be that we forget how rarely we forget.

"Here's the most incredible thing about the test I just gave you," Ed declared. "We could play this game several years from now and ask you which of these photos you've seen before, and you'd actually be able to point to the right one more often than not. Somewhere in your mind there's a trace from everything you've ever seen."

That sounded like a bold and possibly dubious claim, one that I was curious to look into. Exactly how good *are* our memories? I wondered. Is it possible we have the capacity to remember *everything*?

This notion that our brains don't ever really forget is certainly embedded in the way we talk about our memories. The metaphors we most often use to describe memory—the photograph, the tape recorder, the mirror, the computer—all suggest mechanical accuracy, as if the mind were some sort of meticulous transcriber of our experiences. Indeed, I learned that until fairly recently, most psychologists suspected that our brains really do function as perfect recorders—that a lifetime of memories are socked away somewhere in the cerebral attic, and if they can't be found it isn't because they've vanished, but only because we've misplaced them. In an oft-cited paper published in 1980, the psychologist Elizabeth Loftus polled her colleagues and found that fully 84 percent of them agreed with this statement: "Everything we learn is permanently stored in the mind, although sometimes particular details are not accessible. With hypnosis, or other special techniques, these inaccessible details could eventually be recovered."

Loftus goes on to say that this conviction has its modern origins in a set of experiments carried out from 1934 to 1954 by a Canadian neurosurgeon named Wilder Penfield. Penfield used electrical probes to stimulate the brains of epileptic patients while they were lying conscious on the operating table with their skulls exposed. He was trying to pinpoint the source of their epilepsy, and hopefully cure it, but he found that when his probe touched certain parts of his patients' temporal lobes, something very unexpected happened. The patients started describing vivid, long-forgotten memories. When he touched the same spot again, he often elicited the same memory. Based on those experiments, Penfield came to believe that the brain records everything to which it pays any degree of conscious attention, and that this recording is permanent.

The Dutch psychologist Willem Wagenaar came to believe the same thing. For six years, between 1978 and 1984, he kept a diary of the one or two most notable events that happened to him each day. For each event, he wrote down what occurred, who was involved, where it occurred, and when—each on a separate card. In 1984, he began testing himself to see just how much of those six years he'd be able to recall. He would pull out a random card and see if he had any memories of the events described that day. He found that he could recall almost everything that happened—especially the more recent events—with just a few retrieval clues. But nearly 20 percent of the oldest memories seemed to have totally disappeared. These events, described in his own diary, felt totally foreign, as if they had happened to a stranger.

But were those memories really gone? Wagenaar wasn't convinced they were. He decided to take another look at ten events that he believed he'd completely forgotten, in which his diary suggested that another person had been present. He went back to those people and asked them for details that might help him recall his lost memories. In every single case, with enough prodding, someone was able to supply a detail that led Wagenaar to retrieve other parts of the memory. Not one of his memories had actually disappeared. He concluded that "in light of this one cannot say that any event was completely forgotten."

Even so, over the last three decades, most psychologists have grown less optimistic that we in fact possess perfect memories of the past, just waiting to be uncovered. As neuroscientists have begun to unravel some of the mysteries of what exactly a memory is, it's become clear that the fading, mutating, and eventual disappearance of memories over time is a real physical phenomenon that happens in the brain at the cellular level. And most now agree that Penfield's experiments elicited hallucinations—something more like déjà vu or a dream than real memories.

Nevertheless, the sudden reappearance of long-lost episodes from one's past is a familiar enough experience, and the notion that with just the right cue, we might somehow be able to pull out every single bit of information that once went into our brains persists. In fact, probably the single most common misperception about human memory—the one that Ed had so casually laughed off—is that some people have photographic memories. When I followed up with him about that, he confided that he used to wake up in cold sweats worrying that someday someone with a photographic memory would read about the World Memory Championship in the newspaper, show up, and blow him and his colleagues out of the water. He was reassured to learn that most scientists now agree that this is unlikely to happen. Even though many people claim to have a photographic memory, there's no evidence that anyone can actually store mental snapshots and recall them with perfect fidelity. Indeed, only one case of photographic memory has ever been described in the scientific literature.

In 1970, a Harvard vision scientist named Charles Stromeyer III published a paper in *Nature*, one of the world's most respected scientific journals, about a young woman named Elizabeth, a Harvard student, who could perform an astonishing feat. Stromeyer showed Elizabeth's right eye a pattern of ten thousand random dots, and a day later he showed her left eye another dot pattern. Astoundingly, Elizabeth was able to mentally fuse the two images, as if they were one of those "Magic Eye" random dot stereograms that were a fad in the 1990s. When she did, she claimed to see a single, new image where the two dot patterns overlapped. Elizabeth seemed to offer the first conclusive proof that photographic memory is possible. But then, in a soap opera twist, Stromeyer married her, and she was never the subject of further testing.

In 1979, another researcher named John Merritt decided to investigate Stromeyer's claims. He placed a photographic memory test in magazines and newspapers around the country. It consisted of two random dot drawings. Merritt hoped someone might come forward with abilities similar to Elizabeth's and prove that her case was not unique. He figures that roughly one million people tried their hand at the test. Of that number, thirty wrote in with the right answer, and fifteen agreed to be studied by Merritt. But with scientists looking over their shoulders, none of them could pull off Elizabeth's nifty trick.

There are so many unlikely circumstances surrounding the Elizabeth case—the marriage between subject and scientist, the lack of further testing, the inability to find anyone else with her abilities—that some psychologists have concluded that there's something fishy about Stromeyer's findings. He denies it. "We don't have any doubt about our data," he told me over the phone. Still, his one-woman study, he admits, "is not strong evidence for other people having photographic memory."

Growing up, I'd been enchanted by stories about ultra-Orthodox Jews who had memorized all 5,422 pages of the Babylonian Talmud so thoroughly that when a pin was stuck through any of the Talmud's sixty-three tractates, or books, they could tell you which words it passed through on every page. I'd always assumed those stories had to be apocryphal, a bit of Hebrew school lore like the levitating rabbi or the wallet-cum-suitcase made out of foreskins. But as it turns out, the pin-prick Talmudists are as legit members of the Jewish pantheon as the Mighty Atom. In 1917, a psychologist named George Stratton wrote up a study in the journal *Psychological Review* about a group of Polish Talmudic scholars known as the Shass Pollak (literally, the "Talmud Pole") who lived up to their reputation of pinpoint precision. But as he noted in his commentary, despite the impressive memories of the Shass Pollak, "none of them ever attained any prominence in the scholarly world." The Shass Pollak didn't possess photographic memories so much as single-minded perseverance in their studies. If the average person decided he was going to dedicate his entire life to memorizing 5,422 pages of text, he'd also eventually get to be pretty good at it.

So if photographic memory is just a myth, what about the Russian journalist S? If he wasn't taking snapshots in his mind, what exactly was he doing?

S's exceptional memory wasn't the only strange feature of his brain. He also suffered from a rare perceptual disorder known as synesthesia, which caused his senses to be bizarrely intertwined. Every sound S heard had its own color, texture, and sometimes even taste, and evoked "a whole complex of feelings." Some words were "smooth and white," others "as orange and sharp as arrows." The voice of

Luria's colleague, the famous psychologist Lev Vygotsky, was "crumbly yellow." The cinematographer Sergei Eisenstein's voice resembled a "flame with fibres protruding from it."

Words set S's mind ablaze with mental imagery. When you or I hear someone mention the word "elephant" or read the word on this page, we understand immediately that the referent is a large, gray pachyderm with thick legs and an oversize proboscis. But under most circumstances we don't actually conjure up an image of an elephant in our mind's eye. We might, if we choose to, but it takes a little extra effort, and in the course of normal conversation or reading, there's usually no point to it. But that's exactly what S did, automatically and instantaneously, with every word he heard. He couldn't help it. "When I hear the word green, a green flowerpot appears; with the word red I see a man in a red shirt coming toward me; as for blue, this means an image of someone waving a small blue flag from a window," he told Luria. Because every word summoned up an accompanying synesthetic image—sometimes also a taste or smell—S lived in a kind of waking dream, once removed from reality. While one universe unfolded around him, another universe of images blossomed in his mind's eye.

These images that populated S's head were so powerful that they felt at times indistinguishable from reality. "Indeed, one would be hard put to say which was more real for him: the world of imagination in which he lived, or the world of reality in which he was but a temporary guest," Luria wrote. All S had to do was imagine himself running after a train to make his pulse race, or envision sticking his hand in a hot oven to make his temperature rise. He claimed even to be able to abolish pain with his images: "Let's say I'm going to the dentist . . . I sit there and when the pain starts I feel it . . . it's a tiny, orange-red thread. I'm upset because I know that if this keeps up the thread will widen until it turns into a dense mass . . . So I cut the thread, make it smaller and smaller, until it's just a tiny point. And the pain disappears."

Even numbers had their own personalities for S: "Take the number 1. This is a proud, well-built man; 2 is a high-spirited woman; 3 a gloomy person (why, I don't know); 6 a man with a swollen foot; 7 a man with a mustache; 8 a very stout woman—a sack within a sack. As for the number 87, what I see is a fat woman and a man twirling his mustache." But while numbers were brought to life by S's synesthesia, he had trouble understanding abstract concepts and metaphors. "I can only understand what I can visualize," he explained. Words like "infinity" and "nothing" were beyond his grasp. "Take the word *something* for example. For me this is a dense cloud of steam that has the color of smoke. When I hear the word *nothing*, I also see a cloud, but that one is thinner, completely transparent. And when I try to seize a particle of this *nothing*, I get the most minute particles of *nothing*." S was simply unable to think figuratively. An expression like "weigh one's words" evoked images of scales, not prudence. Poetry was virtually impossible to read, unless it was completely literal. Even simple stories proved difficult to understand because his irrepressible image-making would bog him down as he tried to visualize every word, or else send his brain hurtling off to some other associated image, and some other memory.

All of our memories are, like S's, bound together in a web of associations. This is not merely a metaphor, but a reflection of the brain's physical structure. The three-pound mass balanced atop our spines is made up of somewhere in the neighborhood of 100 billion neurons, each of which can make upwards of five to ten thousand synaptic connections with other neurons. A memory, at the most fundamental physiological level, is a pattern of connections between those neurons. Every sensation that we remember, every thought that we think, transforms our brains by altering the connections within that vast network. By the time you get to the end of this sentence, your brain will have physically changed.

If thinking about the word "coffee" makes you think about the color black and also about breakfast and the taste of bitterness, that's a function of a cascade of electrical impulses rocketing around a real physical pathway inside your brain, which links a set of neurons that encode the concept of coffee with others containing the concepts of blackness, breakfast, and bitterness. That much scientists know. But how exactly a collection of cells could "contain" a memory remains among the deepest conundrums of neuroscience.

For all the advances that have been made in recent decades, it's still the case that no one has ever actually seen a memory in the human brain. Though advances in imaging technology have allowed neuroscientists to grasp much of the basic topography of the brain, and studies of neurons have given us a clear picture of what happens inside and between individual brain cells, science is still relatively clueless about what transpires in the circuitry of the cortex, the wrinkled outer layer of the brain that allows us to plan into the future, do long division, and write poetry, and which holds most of our memories. In our knowledge of the brain, we're like someone looking down on a city from a high-flying airplane. We can tell where the industrial and residential neighborhoods are, where the airport is, the locations of the main traffic arteries, where the suburbs begin. We also know, in great detail, what the individual units of the city (citizens, and in this metaphor, neurons) look like. But, for the most part, we can't say where people go when they get hungry, how people make a living, or what any given person's commute looks like. The brain makes sense up close and from far away. It's the in-between—the stuff of thought and memory, the language of the brain—that remains a profound mystery.

One thing is clear, however: The nonlinear associative nature of our brains makes it impossible for us to consciously search our memories in an orderly way. A memory only pops directly into consciousness if it is cued by some other thought or perception—some other node in the nearly limitless interconnected web. So when a memory goes missing or a name gets caught on the tip of the tongue, hunting it down can be frustrating and often futile. We have to stumble in the dark with a flashlight for cues that might lead us back to the piece of information we're looking for—*Her name begins with an L . . . She's a painter . . . I met her at that party a couple years ago*—until one of those other memories calls to mind the one we're looking for. *Ah yes, her name was Lisa!* Because our memories don't follow any kind of linear logic, we can neither sequentially search them nor browse them.

But S could. S's memories were as regimentally ordered as a card catalog. Each piece of information he memorized was assigned its own address inside his brain.

Let's say I asked you to memorize the following list of words: "bear," "truck," "college," "shoe," "drama," "garbage," and "watermelon." You might very well be able to remember all seven of those words, but it's less likely you'd be able to remember them in order. Not so with S. For S, the first piece of information in a list was always, and without fail, inextricably linked to the second piece of information, which could only be followed by the third. It didn't matter whether he was memorizing Dante's *Divine Comedy* or mathematical equations; his memories were always stored in linear chains. Which is why he could recite poems just as easily backward as forward.

S kept his memories rigidly organized by mapping them onto structures and places he already knew well. "When S read through a long series of words, each word would elicit a graphic image. And since the series was fairly long, he had to find some way of distributing these images of his in a mental row or sequence," wrote Luria. "Most often . . . he would 'distribute' them along some roadway or street he visualized in his mind."

When he wanted to commit something to memory, S would simply take a mental stroll down Gorky Street in Moscow, or his home in Torzhok, or some other place he'd once visited, and install each of his images at a different point along the walk. One image might be placed at the doorway of a house, another near a streetlamp, another on top of a picket fence, another in a garden, another on the ledge of a store window. All this happened in his mind's eye as effortlessly as if he were placing real objects along a real street. If asked to memorize those same seven words—"bear," "truck," "college," "shoe," "drama," "garbage," and "watermelon"—he would conjure up an image associated with each of them, and scatter them along one of his many mental paths.

When S wanted to recall the information a day, month, year, or decade later, all he would have to do was rewalk the path where that particular set of memories was stored, and he would see each image in the precise spot where he originally left it. When S did, on rare occasions, forget something, "these omissions . . . were not *defects of memory* but were, in fact, *defects of perception*," wrote Luria. In one instance, S forgot the word "pencil" amid a long list of words that he was supposed to have memorized. Here's his own description of how he forgot it: "I put the image of the pencil near a fence . . . the one down the street, you know. But what happened was that the image fused with that of the fence and I walked right on past without noticing it." On another occasion, he forgot the word "egg." "I had put it up against a white wall and it blended in with the background," he explained.

S's memory was a beast that indiscriminately gobbled up everything it was fed, and had trouble disgorging those pieces of information that were too trivial to be worth keeping. The greatest challenge S faced was learning what Luria called "the art of forgetting." The rich images that every sensation created proved frustratingly indelible. He experimented with different techniques to wipe them from his mind. He tried writing things down, with the hope that he would then no longer feel a need to remember them. When that didn't work, he tried burning the pieces of paper, but he could still see numbers hovering among the embers. Eventually he had an epiphany. One evening, while feeling particularly pestered by a chart of numbers he had earlier memorized, he figured out the secret of forgetting. All he had to do was convince himself that the information he wanted to forget was meaningless. "If I don't want the chart to show up it won't," he exclaimed. "And all it took was for me to realize this!"

One might assume that S's vacuum-cleaner memory would have made him a formidable journalist. I imagined if I could only take notes without taking notes and have at my fingertips every fact I'd ever digested, I'd be immensely better at my job. I'd be better at everything.

But professionally S was a failure. His newspaper gig didn't last long, and he was never able to hold down a steady job. He was, in Luria's estimation, "a somewhat anchorless person, living with the expectation that at any moment something particularly fine was to come his way." Ultimately, his condition made him unemployable as anything but a stage performer, a theatrical curio like the mnemonist of Alfred Hitchcock's *The 39 Steps*. The man with the best memory in the world simply remembered too much.

In his short story "Funes the Memorious," Jorge Luis Borges describes a fictional version of S, a man with an infallible memory who is crippled by an inability to forget. He can't distinguish between the trivial and the important. Borges's character Funes can't prioritize, can't generalize. He is "virtually incapable of general, platonic ideas." Like S, his memory was too good. Perhaps, as Borges concludes in his story, it is forgetting, not remembering, that is the essence of what makes us human. To make sense of the world, we must filter it. "To think," Borges writes, "is to forget."

While S's capacious memory for facts seems almost unbelievable, he was in fact taking advantage of the well-developed spatial memory we all possess. If you visit London, you'll occasionally cross paths with young men (and less often women) on motor scooters, blithely darting in and out of traffic while studying maps affixed to their handlebars. These studious cyclists are training to become London cabdrivers. Before they can receive accreditation from London's Public Carriage Office, cabbies-in-training must spend two to four years memorizing the locations and traffic patterns of all 25,000 streets in the vast and vastly confusing city, as well as the locations of 1,400 landmarks. Their training culminates in an infamously daunting exam called "the Knowledge," in which they not only have to plot the shortest route between any two points in the metropolitan area, but also name important places of interest along the way. Only about three out of ten people who train for the Knowledge obtain certification.

In 2000, a neuroscientist at University College London named Eleanor Maguire wanted to find out what effect, if any, all that driving around the labyrinthine streets of London might have on the cabbies' brains. When she brought sixteen taxi drivers into her lab and examined their brains in an MRI scanner, she found one surprising and important difference. The right posterior hippocampus, a part of the brain known to be involved in spatial navigation, was 7 percent larger than normal in the cabbies—a small but very significant difference. Maguire concluded that all of that way-finding around London had physically altered the gross structure of their brains. The more years a cabbie had been on the road, the more pronounced the effect.

The brain is a mutable organ, capable—within limits—of reorganizing itself and readapting to new kinds of sensory input, a phenomenon known as neuroplasticity. It had long been thought that the adult brain was incapable of spawning new neurons—that while learning caused synapses to rearrange themselves and new links between brain cells to form, the brain's basic anatomical structure was more or less static. Maguire's study suggested the old inherited wisdom was simply not true.

After her groundbreaking study of London cabbies, Maguire decided to turn her attention to mental athletes. She teamed up with Elizabeth

Valentine and John Wilding, authors of the academic monograph *Superior Memory*, to study ten individuals who had finished near the top of the World Memory Championship. They wanted to find out if the memorizers' brains were—like the London cabbies'—structurally different from the rest of ours, or if they were somehow just making better use of memory abilities that we all possess.

The researchers put both the mental athletes and a group of matched control subjects into MRI scanners and asked them to memorize three-digit numbers, black-and-white photographs of people's faces, and magnified images of snowflakes, while their brains were being scanned. Maguire and her team thought it was possible that they might discover anatomical differences in the brains of the memory champs, evidence that their brains had somehow reorganized themselves in the process of doing all that intensive remembering. But when the researchers reviewed the imaging data, not a single significant structural difference turned up. The brains of the mental athletes appeared to be indistinguishable from those of the control subjects. What's more, on every single test of general cognitive ability, the mental athletes' scores came back well within the normal range. The memory champs weren't smarter, and they didn't have special brains. When Ed and Lukas told me they were average guys with average memories, they weren't just being modest.

But there was one telling difference between the brains of the mental athletes and the control subjects: When the researchers looked at which parts of the brain were lighting up when the mental athletes were memorizing, they found that they were activating entirely different circuitry. According to the functional MRIs, regions of the brain that were less active in the control subjects seemed to be working in overdrive for the mental athletes.

Surprisingly, when the mental athletes were learning new information, they were engaging several regions of the brain known to be involved in two specific tasks: visual memory and spatial navigation, including the same right posterior hippocampal region that the London cabbies had enlarged with all their daily way-finding. At first glance, this wouldn't seem to make any sense. Why would mental athletes be conjuring images in their mind's eye when they were trying to learn three-digit numbers? Why should they be navigating like London cabbies when they're supposed to be remembering the shapes of snowflakes?

Maguire and her team asked the mental athletes to describe exactly what was going through their minds as they memorized. The mental athletes recounted a strategy that sounded almost exactly like what S claimed had been happening in his brain. Even though they were not innate synesthetes like S, the mental athletes said they were consciously converting the information they were being asked to memorize into images, and distributing those images along familiar spatial journeys. Unlike S, they weren't doing this automatically, or because it was an inborn talent they'd nurtured since childhood. Rather, the unexpected patterns of neural activity that Maguire's fMRIs turned up were the result of training and practice. The mental athletes had taught themselves to remember like S.

I found myself fascinated by Ed and his quiet friend Lukas, and this formidable-sounding project of theirs to push their memories as hard and as far as they possibly could. And they likewise seemed fascinated with me, a journalist of roughly the same age, who might share their story in some magazine they'd never heard of, and perhaps jump-start their careers as mnemonic celebrities. After Ed's lecture at the high school, he invited me to follow him and Lukas to a nearby bar, where we met up with an aspiring filmmaker and old boarding school chum of Ed's who had been trailing them around New York with an

8-mm video camera, documenting their every antic adventure, including Lukas's attempt to memorize a deck of playing cards on the fifty-three-second elevator ride to the Empire State Building's observation deck. ("We wanted to see if the fastest lift in the world was faster than the Austrian speed cards champion," Ed deadpanned. "It wasn't.")

After a few drinks, Ed was keen to carry me deeper into the obscure underworld of mental athletic secrets. He offered to introduce me to the rituals of the KL7, a "secret society of memorizers" that he and Lukas cofounded at the Kuala Lumpur championships in 2003, and which, evidently, was not so secret.

"KL, as in Kuala Lumpur?" I asked.

"No, KL as in Knights of Learning, and the seven is because it started with seven of us," Lukas explained, while sipping one of the three free beers he had just won by memorizing a deck of cards for the waitress. "It's an international society for the development of education."

"Membership in our society is an extraordinarily high honor," Ed added.

Though the club's endowment of more than a thousand dollars languishes in Lukas's bank account, Ed conceded that the KL7 has never actually done much of anything, except get drunk together the evening after memory contests (occasionally aided by a sophisticated pressurized keg attachment designed by Lukas that folds up into a suitcase). When I pressed Ed for more information, he offered to demonstrate the society's single cherished ceremony.

"Just call it a satanic ritual," he said, and then asked Jonny, his documentarian, to set a timer on his wristwatch. "We each have exactly five minutes to drink two beers, kiss three women, and memorize forty-nine random digits. Why forty-nine digits? It's seven squared."

"I was surprised to discover that this is actually quite difficult," said Lukas. He was wearing a shiny charcoal suit and a shinier tie, and had no trouble convincing the waitress, whom he'd already won over, to give him three pecks on the cheek.

"Technically that's unsatisfactory, but we'll count it," Ed proclaimed, a rivulet of beer running down his chin. From his pocket he pulled out a page of printed numbers and tore it into sections. His finger raced across the scrap until it got to the forty-ninth digit, at which point he stood up and sputtered, "Almost done!" and then limped over to a nearby booth, where he tried to explain his predicament to three silver-haired women who seemed far too old to be enjoying this loud bar. With time running out, and before they could respond to his plea, he had leaned across the table and planted his lips on each of their sunken, flustered cheeks.

Ed returned triumphantly, pumping his fist and soliciting high fives from all of us. He ordered another round for the table.

I didn't know quite what to make of Ed. He was, I was gradually discovering, an aesthete, in the true Oscar Wilde sense. More than anyone I'd ever met, he seemed to participate in life as if it were art, and to practice a studied, careful carefreeness. His sense of what is worthy seemed to overlap very little with any conventional sense of what is useful, and if there were one precept that could be said to govern his life, it is that one's highest calling is to engage in enriching escapades at every turn. He was a genuine bon vivant, and yet he approached the subject of his PhD research, the relationship between memory and perception, with a rigor and seriousness that suggested he intended to accomplish big things. He was in no conventional sense handsome, and yet later that night, I watched him approach a woman in the street, ask for a cigarette, and a few minutes later walk away reciting her phone number. His "normal bar trick," he told me, involves shimmying up to a young lady and inviting her to create an "arbitrarily long number," and then promising to buy her a bottle of champagne should he successfully remember it.

Over the course of the evening, Ed regaled me with story after story of his adventures and instructive misadventures. There was the time he threw his shoeless self through the window of a bar in New Zealand in order to circumvent a bouncer. The time he crashed a supermodel's party in London. ("It was easier then, I was in a wheelchair, and I could do a superior wheelie.") The time he crashed a party at the British embassy in Paris. ("I noticed the ambassador following my dirty shoes all the way across the room.") And how could he forget the twelve hours he spent panhandling for bus fare in downtown Los Angeles?

At the time, I may have sounded a note of skepticism about these self-mythologizing stories, but that was only because I didn't yet know Ed well enough to recognize that he very well could have been understating their outrageousness. A few more drinks into the evening, it dawned on me that I'd spent the better part of the day with Ed and Lukas and neither of them had once called me by my name, though I was sure I'd told it to them when I first introduced myself. Ed had referred to me in front of the waitress as "our journalist friend," and Lukas just hadn't referred to me. These were evasions I knew well. But Ed had assured me earlier in the day that he could memorize the name and phone number of every girl he ever met. I thought that sounded like the kind of impressive skill that was bound to take one far in life. Bill Clinton is supposed to never forget a name and, well, look where that got him. But it occurred to me now that Ed's "could" was a bit ambiguous, and might have been of the same nature as "He could count backward from a million"—as in, yeah, if he really wanted to. I asked Ed if he remembered my name.

"Of course. It's Josh."

"My last name?"

"Shit. Did you tell it to me?"

"Yes, Foer. Josh Foer. You're human after all."

"Ah, well—"

"I thought you were supposed to have a fancy technique for remembering people's names."

"In theory, I do. But its utility is inversely proportional to the amount of alcohol I've imbibed."

Ed then explained to me his procedure for making a name memorable, which he had used in the competition to memorize the first and last names associated with ninety-nine different photographic head shots in the names-and-faces event. It was a technique he promised I could use to remember people's names at parties and meetings. "The trick is actually deceptively simple," he said. "It is always to associate the sound of a person's name with something you can clearly imagine. It's all about creating a vivid image in your mind that anchors your visual memory of the person's face to a visual memory connected to the person's name. When you need to reach back and remember the person's name at some later date, the image you created will simply pop back into your mind . . . So, hmm, you said your name was Josh Foer, eh?" He raised an eyebrow and gave his chin a melodramatic stroke. "Well, I'd imagine you joshing me where we first met, outside the competition hall, and I'd imagine myself breaking into four pieces in response. Four/Foer, get it? That little image is more entertaining—to me, at least—than your mere name, and should stick nicely in the mind." It occurred to me that this was a kind of manufactured synesthesia.

To understand why this sort of mnemonic trick works, you need to know something about a strange kind of forgetfulness that psychologists have dubbed the "Baker/baker paradox." The paradox goes like this: A researcher shows two people the same photograph of a face and tells one of them that the guy is a baker and the other that his last name is Baker. A couple days later, the researcher shows the same two guys the same photograph and asks for the accompanying word. The person who was told the man's profession is much more likely to remember it than the person who was given his surname. Why should that be? Same photograph. Same word. Different amount of remembering.

When you hear that the man in the photo is a baker, that fact gets embedded in a whole network of ideas about what it means to be a baker: He cooks bread, he wears a big white hat, he smells good when he comes home from work. The name Baker, on the other hand, is tethered only to a memory of the person's face. That link is tenuous, and should it dissolve, the name will float off irretrievably into the netherworld of lost memories. (When a word feels like it's stuck on the tip of the tongue, it's likely because we're accessing only part of the neural network that "contains" the idea, but not all of it.) But when it comes to the man's profession, there are multiple strings to reel the memory back in. Even if you don't at first remember that the man is a baker, perhaps you get some vague sense of breadiness about him, or see some association between his face and a big white hat, or maybe you conjure up a memory of your own neighborhood bakery. There are any number of knots in that tangle of associations that can be traced back to his profession. The secret to success in the names-and-faces event—and to remembering people's names in the real world—is simply to turn Bakers into bakers—or Foers into fours. Or Reagans into ray guns. It's a simple trick, but highly effective.

I tried using the technique myself to remember the name of the documentary filmmaker who had been trailing Ed and Lukas around town all week. He introduced himself as Jonny Lowndes. "We call him Pounds Lowndes," Ed interjected. "He used to be heavyset in high school." Since my older brother's childhood nickname was Jonny, I closed my eyes and pictured the two of them together, arm in arm, gobbling up a pound cake.

"You know we could teach you more tricks like that," Ed said. He turned to Lukas ebulliently. "I'm trying to think if by the end of the night we could have him winning the American championship?"

"I get the sense that you hold the Americans in rather low esteem," I said.

"On the contrary, they just haven't had the right coach," he said, turning back to me. "I reckon you could win the championship next year with an hour's practice a day." He looked to Lukas. "Don't you think that's right?"

Lukas nodded.

"You and Tony Buzan both," I said.

"Ah, yes, the estimable Tony Buzan," Ed scoffed. "Did he try to sell you that nonsense about the brain being a muscle?"

"Um, yes, he did."

"Anyone who knows the first thing about the respective characteristics of brains and muscles knows how risible that analogy is." It was my first hint of Ed's tortured relationship with Buzan. "Look, what you really need to do is bring me on as your coach, trainer, and manager—and, um, spiritual yogi."

"And what would you get out of this relationship?" I asked.

"I'd get pleasure," he responded with a smile. "Also, you being a journalist, I wouldn't mind if, in the course of your writing about this experience, you managed to give the impression that I would be an excellent person to have tutoring your daughter in the Hamptons at, like, a squillion quid an hour."

I laughed and told Ed that I'd give it some thought. I honestly wasn't that interested in spending an hour a day pawing through playing cards, or memorizing pages of random numbers, or doing any of the other mental calisthenics that seemed to be involved in becoming a "mental athlete." I have always embraced my own nerdiness—I was captain of my high school quiz bowl team and have long worn a watch with calculator functions—but this was a bit much even for me. And yet I was curious enough about learning where the limits of my memory lay, and intrigued enough by Ed, to consider this exercise.

All the mental athletes I'd met had insisted that anyone was capable of improving his or her memory—that the untapped powers of S are inside all of us. I decided I was going to try to find out if that was really true. That night, when I got home, there was a short e-mail from Ed waiting in my in-box: "So, anyway, can I be your coach?"

THREE

THE EXPERT EXPERT

Though it's best not to be born a chicken at all, it is especially bad luck to be born a cockerel.

From the perspective of the poultry farmer, male chickens are useless. They can't lay eggs, their meat is stringy, and they're ornery to the hens that do all the hard work of putting food on our tables. Commercial hatcheries tend to treat male chicks like fabric cutoffs or scrap metal: the wasteful but necessary by-product of an industrial process. The sooner they can be disposed of—often they're ground into animal feed—the better. But a costly problem has vexed egg farmers for millennia: It's virtually impossible to tell the difference between male and female chickens until they're four to six weeks old, when they begin to grow distinctive feathers and secondary sex characteristics like the rooster's comb. Until then, they're all just indistinguishable fluff balls that have to be housed and fed—at considerable expense.

Somehow it took until the 1920s before anyone figured out a solution to this costly dilemma. The momentous discovery was made by a group of Japanese veterinary scientists, who realized that just inside the chick's rear end there is a constellation of folds, marks, spots, and bumps that to the untrained eye appear arbitrary, but when properly read, can divulge the sex of a day-old bird. When this discovery was unveiled at the 1927 World Poultry Congress in Ottawa, it revolutionized the global hatchery industry and eventually lowered the price of eggs worldwide. The professional chicken sexer, equipped with a skill that took years to master, became one of the most valuable workers in agriculture. The best of the best were graduates of the two-year Zen-Nippon Chick Sexing School, whose standards were so rigorous that only 5 to 10 percent of students received accreditation. But those who did graduate earned as much as five hundred dollars a day and were shuttled around the world from hatchery to hatchery like top-flight business consultants. A diaspora of Japanese chicken sexers spilled across the globe.

Chicken sexing is a delicate art, requiring Zen-like concentration and a brain surgeon's dexterity. The bird is cradled in the left hand and given a gentle squeeze that causes it to evacuate its intestines (too tight and the intestines will turn inside out, killing the bird and rendering its gender irrelevant). With his thumb and forefinger, the sexer flips the bird over and parts a small flap on its hindquarters to expose the cloaca, a tiny vent where both the genitals and anus are situated, and peers deep inside. To do this properly, his fingernails have to be precisely trimmed. In the simple cases—the ones that the sexer can actually explain—he's looking for a barely perceptible protuberance called the "bead," about the size of a pinhead. If the bead is convex, the bird is a boy, and gets thrown to the left; concave or flat and it's a girl, sent down a chute to the right. Those cases are easy enough. In fact, a study has shown that amateurs can be taught to identify the bead with only a few minutes of training. But in roughly 80 percent of the chicks, the bead is not obvious and there is no single distinguishing trait the sexer can point to.

By some estimates there are as many as a thousand different vent configurations that a sexer has to learn to become competent. The job is made even more difficult by the fact that the sexer has to diagnose the bird with just a glance. There is no time for conscious reasoning. If he hesitates for even a couple seconds, his grip on the bird can cause a pullet's vent to swell to look unquestionably like a cockerel's. Mistakes are costly. In the 1960s, one hatchery paid its sexers a penny for each correctly sexed chick and deducted 35 cents for each one they got wrong. The best in the business can sex 1,200 chicks an hour with 98 to 99 percent accuracy. In Japan, a few superheroes of the industry have learned how to double clutch the chicks and sex them two at a time, at the rate of 1,700 per hour.

What makes chicken sexing such a captivating subject—the reason that academic philosophers and cognitive psychologists have authored dissertations about it, and the reason that my own research into memory had brought me to this arcane skill—is that even the best professional sexers can't describe how they determine gender in the toughest, most ambiguous cases. Their art is inexplicable. They say that within three seconds they just "know" whether a bird is a boy or a girl, but they can't say how they know. Even when carefully cross-examined by researchers, they can't give reasons why one bird is a male and another is female. What they have, they say, is intuition. In some fundamental sense, the expert chicken sexer perceives the world—at least the world of chicken privates—in a way that is completely different from you or me. When they look at a chick's bottom, they see things that a normal person simply does not see. What does chicken sexing have to do with my memory? Everything.

. . .

I decided it would be a good idea to dive (bellyflop, really) into the scientific literature. I was looking for some hard evidence that our memories might really be improvable in the dramatic way that Buzan and the mental athletes had promised. I didn't have to search very hard. As I was combing the scientific literature, one name kept popping up in my research about memory improvement: K. Anders Ericsson. He was a psychology professor at Florida State University and the author of an article titled "Exceptional Memorizers: Made, Not Born."

Before Tony Buzan mass-marketed the idea of "using your perfect memory," Ericsson was laying the scientific groundwork for what's known as "Skilled Memory Theory," which explains how and why our memory is improvable. In 1981, he and fellow psychologist Bill Chase conducted a now-classic experiment on a Carnegie Mellon undergraduate, who has been immortalized in the literature by his initials, SF. Chase and Ericsson paid SF to spend several hours a week in their lab taking a simple memory test over and over and over again. It was similar to the test that Luria had given to S when he first walked into his office. SF sat in a chair and tried to remember as many numbers as possible as they were read off at the rate of one per second. At the outset, he could only hold about seven digits in his head at a time. By the time the experiment wrapped up—two years and 250 mind-numbing hours later—SF had expanded his ability to remember numbers by a factor of ten. The experiment shattered the old notions that our memory capacities are fixed. How SF did it, Ericsson believes, holds a key to understanding the basic cognitive processes underlying all forms of expertise—from mental athlete memorizers to chess grand masters to chicken sexers.

Everyone has a great memory for something. We've already seen the mnemonic gifts of London cabbies, and the scientific literature is filled with papers about the "superior memories" of waiters, the vast capacities of actors to remember lines, and the memory skills possessed by experts in a wide variety of other fields. Researchers have studied the exceptional memories of doctors, baseball fans, violinists, soccer players, snooker players, ballet dancers, abacus wranglers, crossword puzzlers, and volleyball defenders. Pick any human endeavor in which people excel, and I'll give you even odds that some psychologist somewhere has written a paper about the exceptional memories possessed by experts in that field.

Why is it that veteran waiters don't have to write down orders? Why are the best violinists in the world so good at memorizing new musical scores? How come, as one study proved, elite soccer players can glance at a soccer match on TV and reconstruct almost exactly what was happening in the game? One possible explanation might be that people with good memories for dinner orders get channeled into the food-service industry, or that the soccer players with the best memory for arrangements of players have a knack for clawing their way up to the premier league, or that people with a great eye for chicken ass naturally gravitate to the Zen-Nippon Chick Sexing School. But that seems unlikely. It makes much more sense to believe the causality works in the opposite direction. There is something about mastering a specific field that breeds a better memory for the details of that field. But what is that something? And can that something somehow be generalized, so that anyone can acquire it?

The Human Performance Lab, which Ericsson runs with a group of other FSU researchers, is where experts come to have their memories—and much else—tested. Ericsson is probably the world's leading expert on experts. Indeed, he has achieved a degree of popular fame in recent years thanks to his research showing that experts tend to require at least ten thousand hours of training to achieve their world-class status. When I called him up and told him that I had been thinking about trying to train my own memory, he wanted to know whether I had started yet. I said I hadn't really begun. He was thrilled; he told me he almost never gets the chance to study a novice in the process of becoming an expert. If I was serious, he said, he wanted to make me his research subject. He invited me down to Florida for a couple days to take a few tests. He wanted to get some baseline measurements of my memory before I started trying to improve it.

The Human Performance Lab occupies a plush office complex on the outskirts of Tallahassee. The bookshelves that line the walls overflow with an eclectic catalog of titles that have been relevant to Ericsson's research: *The Musical Temperament*, *Surgery of the Foot*, *How to Be a Star at Work*, *Secrets of Modern Chess Strategy*, *Lore of Running*, *The Specialist Chick Sexer*.

David Rodrick, a young research associate in the lab, gleefully described the place as "our toy palace." When I arrived a couple weeks after my initial phone call with Ericsson, there was a floor-to-ceiling nine-by-fourteen-foot screen set up in the middle of one of the rooms displaying life-size video footage of a traffic stop. It was shot from the perspective of a police officer walking up to a stopped car.

For the previous few weeks, Ericsson and his colleagues had been bringing members of the Tallahassee SWAT team and recent graduates of the police academy into his lab and placing them in front of the big screen with a Beretta handgun loaded with blanks holstered to their belt. They bombarded the officers with one hair-raising scenario after another and watched how they responded. In one scenario, the officer saw a man walk toward the front door of a school with a suspicious bulge that looked like a bomb strapped to his chest. The researchers wanted to know how officers with different levels of experience would react.

The results were striking. Experienced SWAT officers immediately pulled their guns and yelled repeatedly for the suspect to stop. When he didn't, they almost always shot him before he made it into the school. But recent graduates of the academy were more likely to let the man with the bomb stroll right up the steps and into the building. They simply lacked the experience to diagnose the situation and react properly. At least that would be the superficial explanation. But what exactly does experience mean? What exactly did the more senior officers see that the younger recruits didn't? What were they doing with their eyes, what was going through their minds, how were they processing the situation differently? What were they pulling from their memories? Like the professional chicken sexers, the senior SWAT officers had a skill that was difficult to put into words. Ericsson's research program can be summarized as an attempt to isolate the thing we call expertise, so that he can dissect it and identify its cognitive basis.

In order to do that, Ericsson and his colleagues asked the officers to talk aloud about what was going through their minds as the scenario unfolded. What Ericsson expected to learn from these accounts was the same thing he's found in every other field of expertise that he's studied: Experts see the world differently. They notice things that nonexperts don't see. They home in on the information that matters most, and have an almost automatic sense of what to do with it. And most important, experts process the enormous amounts of information flowing through their senses in more sophisticated ways. They can overcome one of the brain's most fundamental constraints: the magical number seven.

. . .

In 1956, a Harvard psychologist named George Miller published what would become a classic paper in the history of memory research. It began with a memorable introduction:

> My problem is that I have been persecuted by an integer. For seven years this number has followed me around, has intruded in my most private data, and has assaulted me from the pages of our most public journals. This number assumes a variety of disguises, being sometimes a little larger and sometimes a little smaller than usual, but never changing so much as to be unrecognizable. The persistence with which this number plagues me is far more than a random accident. There is, to quote a famous senator, a design behind it, some pattern governing its appearances. Either there really is something unusual about the number or else I am suffering from delusions of persecution.

In fact, we are all persecuted by the integer Miller was referring to. His paper was titled "The Magical Number Seven, Plus or Minus Two: Some Limits on Our Capacity for Processing Information." Miller had discovered that our ability to process information and make decisions in the world is limited by a fundamental constraint: We can only think about roughly seven things at a time.

When a new thought or perception enters our head, it doesn't immediately get stashed away in long-term memory. Rather, it exists in a temporary limbo, in what's known as working memory, a collection of brain systems that hold on to whatever is rattling around in our consciousness at the present moment.

Without looking back and rereading it, try to repeat the first three words of this sentence to yourself.

Without looking back

Easy enough.

Now, without looking back, try to repeat the first three words of the sentence before that. If you find that quite a bit harder, it's because that sentence has already been dropped by your working memory.

Our working memories serve a critical role as a filter between our perception of the world and our long-term memory of it. If every sensation or thought was immediately filed away in the enormous database that is our long-term memory, we'd be drowning, like S and Funes, in irrelevant information. Most of the things that pass through our brain don't need to be remembered any longer than the moment or two we spend perceiving them and, if necessary, reacting to them. In fact, dividing memory between short-term and long-term stores is such a savvy way of managing information that most computers are built around the same model. They have long-term memories in the form of hard drives as well as a working memory cache in the CPU that stores whatever the processor is computing at the moment.

Like a computer, our ability to operate in the world, is limited by the amount of information we can juggle at one time. Unless we repeat things over and over, they tend to slip from our grasp. Everyone knows our working memory stinks. Miller's paper explained that it stinks within very specific parameters. Some people can hold as few as five things in their head at any given time, a few people can hold as many as nine, but the "magical number seven" seems to be the universal carrying capacity of our short-term working memory. To make matters worse, those seven things only stick around for a few seconds, and often not at all if we're distracted. This fundamental limitation, which we all share, is what makes us find the feats of memory gurus so amazing.

My own memory test did not occur in front of the Human Performance Lab's floor-to-ceiling projection screen. There were no guns holstered to my belt, no eye-tracking devices attached to my head. My humble contribution to human knowledge was extracted in Room 218 of the FSU psychology department, a small windowless office with a stained carpet and old IQ tests strewn across the floor. Ungenerously, it might be described as a storage closet.

The man administering my tests was a third-year PhD student in Ericsson's lab named Tres Roring. Though his flip-flops and blond surfer mop might not suggest it, Tres grew up in a small town in southern Oklahoma, where his father is an oil man. At age sixteen, he was the Oklahoma State Junior Chess Champion. His full name is Roy Roring III—hence "Tres."

Tres and I spent three full days in Room 218 taking memory test after memory test—me wearing a clunky microphone headset attached to an old tape recorder, Tres sitting behind me, legs crossed, with a stopwatch in his lap, taking notes.

There were tests of my memory for numbers (forward and backward), tests of my memory for words, tests of my memory for people's faces, and tests of all sorts of things that seemed unlikely to have anything to do with my memory—like whether I could visualize rotating cubes in my mind's eye, and whether I knew the definitions of "jocose," "lissome," and "querulous." Another multiple-choice exam called the

Multidimensional Aptitude Battery Information Test gauged my Trivial Pursuit skills with questions like:

> When did Confucius live?
> A) 1650 A.D.
> B) 1200 A.D.
> C) 500 A.D.
> D) 500 B.C.
> E) 40 B.C.

and:

> In a gasoline engine, the main purpose of the carburetor is to
> A) mix gasoline and air
> B) keep the battery charged
> C) ignite the fuel
> D) contain the pistons
> E) pump the fuel into the engine

Many of the tests Tres administered were lifted directly from USA Memory Championship events, like the fifteen-minute poem, names and faces, random words, speed numbers, and speed cards. He wanted to see how I'd do on them before I'd ever tried to improve my memory. He also wanted to test me on a few of the events that are only used in international memory competitions, like binary digits, historical dates, and spoken numbers. By the end of my three days in Tallahassee, Tres had collected seven hours of audiotaped data for Ericsson and his grad students to analyze later. Lucky them.

And then there were the extensive interviews conducted by another graduate student, Katy Nandagopal. *Do you think you have a good natural memory?* (Pretty good, but nothing special.) *Did you ever play memory games growing up?* (Not that I can think of.) *Board games?* (Only with my grandmother.) *Do you enjoy riddles?* (Who doesn't?) *Can you solve a Rubik's cube?* (No.) *Do you sing?* (Only in the shower.) *Dance?* (Ditto.) *Do you work out?* (Sore subject.) *Do you use workout tapes?* (You need to know that?) *Do you have electrical wiring expertise?* (Really?)

For someone who wants to know what's being done to him so that he might someday tell other people about it, being the subject of a scientific study can be exceedingly trying.

"Why exactly are we doing this?" I'd ask Tres.

"I'd rather not tell you everything right now." (If there was something I was going to be tested on later—and as it turned out, there was—he didn't want me to know.)

"How did I do on that last test?"

"We'll let you know when this is all done."

"Can you at least tell me about your hypothesis?"

"Not now."

"What's my IQ?"

"I don't know."

"High, though?"

The mind-numbing memory exam that SF, the Carnegie Mellon undergraduate, took over and over again for 250 hours for two years is known as the digit span test. It is a standard measure of a person's working-memory capacity for numbers. Most people who are given the test are like SF when he started: They're only able to remember seven plus-or-minus two digits. Most people remember those seven plus-or-minus two numbers by repeating them over and over again to themselves in the "phonological loop," which is just a fancy name for the little voice that we can hear inside our head when we talk to ourselves. The phonological loop acts as an echo, producing a short-term memory buffer that can store sounds just a couple seconds, if we're not rehearsing them. When he began participating in Chase and Ericsson's experiment, SF also used his phonological loop to store information. And for a long time his scores on the test didn't improve. But then something happened. After hours of testing, SF's scores started inching up. One day he remembered ten digits. The next day it was eleven. The number of digits he could recall kept rising steadily. He had made a discovery: Even if his short-term memory was limited, he'd figured out a way to store information directly in long-term memory. It involved a technique called chunking.

Chunking is a way to decrease the number of items you have to remember by increasing the size of each item. Chunking is the reason that phone numbers are broken into two parts plus an area code and that credit card numbers are split into groups of four. And chunking is extremely relevant to the question of why experts so often have such exceptional memories.

The classic explanation of chunking involves language. If you were asked to memorize the twenty-two letters HEADSHOULDERS-KNEESTOES, and you didn't notice what they spelled, you'd almost certainly have a tough time with it. But break up those twenty-two letters into four chunks—HEAD, SHOULDERS, KNEES, and TOES—and the task becomes a whole lot easier. And if you happen to know the full nursery rhyme, the line "Head, shoulders, knees, and toes" can effectively be treated like one single chunk. The same can be done with numbers. The twelve-digit numerical string 120741091101 is pretty hard to remember. Break it into four chunks—120, 741, 091, 101—and it becomes a little easier. Turn it into two chunks, 12/07/41 and 09/11/01, and they're almost impossible to forget. You could even turn those dates into a single chunk of information by remembering it as "the two big surprise attacks on American soil."

Notice that the process of chunking takes seemingly meaningless information and reinterprets it in light of information that is already stored away somewhere in our long-term memory. If you didn't know the dates of Pearl Harbor or September 11, you'd never be able to chunk that twelve-digit numerical string. If you spoke Swahili and not English, the nursery rhyme would remain a jumble of letters. In other words, when it comes to chunking—and to our memory more broadly—what we already know determines what we're able to learn.

Though he'd never been properly taught the technique of chunking, SF figured it out on his own. An avid runner, he began thinking of the strings of random numbers as running times. For example 3,492 was turned into "3 minutes and 49 point 2 seconds, near world-record mile time." And 4,131 became "4 minutes, 13 point 1 seconds, a mile time." SF didn't know anything about the random numbers he had to memorize, but he did know about running. He discovered that he could take meaningless bits of information, run them through a filter that applied meaning to them, and make that information much stickier. He had taken his past experiences and used them to shape how he perceived the present. He was using associations in his long-term memory to see the numbers differently.

This, of course, is what all experts do: They use their memories to see the world differently. Over many years, they build up a bank of experience that shapes how they perceive new information. The experienced SWAT officer doesn't just see a man walking up the front steps of the school; he sees a nervous twitch in the man's arm that calls up associations with dozens of other nervous twitches he's seen in his years of policing. He sees the suspect in the context of every other suspicious person he's ever come across. He perceives the current encounter in light of past encounters like it.

When a graduate of the Zen-Nippon Chick Sexing School looks at a chick's bottom, finely honed perceptual skills allow the sexer to quickly and automatically gather up a stock of information embedded in the chick's anatomy, and before a conscious thought can even enter his or her head, the sexer knows whether the chick is a boy or a girl. But as with the senior SWAT officer, that seemingly automatic knowledge is hard earned. It is said that a student of sexing must work through at least 250,000 chicks before attaining any degree of proficiency. Even if the sexer calls it "intuition," it's been shaped by years of experience. It is the vast memory bank of chick bottoms that allows him or her to recognize patterns in the vents glanced at so quickly. In most cases, the skill is not the result of conscious reasoning, but pattern recognition. It is a feat of perception and memory, not analysis.

The classic example of how memories shape the perception of experts comes from what would seem to be the least intuitive of fields: chess. Practically since the origins of the modern game in the fifteenth century, chess has been regarded as the ultimate test of cognitive ability. In the 1920s, a group of Russian scientists set out to quantify the intellectual advantages of eight of the world's best chess players by giving them a battery of basic cognitive and perceptual tests. To their surprise, the researchers found that the grand masters didn't perform significantly better than average on any of their tests. The greatest chess players in the world didn't seem to possess a single major cognitive advantage.

But if chess masters aren't, as a whole, smarter than lesser chess players, then what are they? In the 1940s, a Dutch psychologist and chess aficionado named Adriaan de Groot asked what seemed like a simple question: What separates merely good chess players from those who are world-class? Did the best-class players see more moves ahead?

Did they ponder more possible moves? Did they have better tools for analyzing those moves? Did they simply have a better intuitive grasp of the dynamics of the game?

One of the reasons chess is such a satisfying game to play and to study is that your average chess buff can be utterly befuddled by a master's move. Often the best move seems entirely counterintuitive. Realizing this, De Groot pored through old games between chess masters and selected a handful of board positions where there was definitely one correct, but not obvious, move to be made. He then presented the boards to a group of international chess masters and top club players. He asked them to think aloud while they brooded over the proper move.

What De Groot uncovered was an even bigger surprise than what his Russian predecessors had found. For the most part, the chess experts didn't look more moves ahead, at least not at first. They didn't even consider more possible moves. Rather, they behaved in a manner surprisingly similar to the chicken sexers: They tended to see the right moves, and they tended to see them almost right away.

It was as if the chess experts weren't thinking so much as reacting. When De Groot listened to their verbal reports, he noticed that they described their thoughts in different language than less experienced chess players. They talked about configurations of pieces like "pawn structures" and immediately noticed things that were out of sorts, like exposed rooks. They weren't seeing the board as thirty-two pieces. They were seeing it as chunks of pieces, and systems of tension.

Grand masters literally see a different board. Studies of their eye movements have found that they look at the edges of squares more than inexperienced players, suggesting that they're absorbing information from multiple squares at once. Their eyes also dart across greater distances, and linger for less time at any one place. They focus on fewer different spots on the board, and those spots are more likely to be relevant to figuring out the right move.

But the most striking finding of all from these early studies of chess experts was their astounding memories. The experts could memorize entire boards after just a brief glance. And they could reconstruct long-ago games from memory. In fact, later studies confirmed that the ability to memorize board positions is one of the best overall indicators of how good a chess player somebody is. And these chess positions are not simply encoded in transient short-term memory. Chess experts can remember positions from games for hours, weeks, even years afterward. Indeed, at a certain point in every chess master's development, keeping mental track of the pieces on the board becomes such a trivial skill that they can take on several opponents at once, entirely in their heads.

As impressive as the chess masters' memories were for chess games, their memories for everything else were notably unimpressive. When the chess experts were shown random arrangements of chess pieces—ones that couldn't possibly have been arrived at through an actual game—their memory for the board was only slightly better than chess novices'. They could rarely remember the positions of more than seven pieces. These were the same chess pieces, and the same chessboards. So why were they suddenly limited by the magical number seven?

The chess experiments reveal a telling fact about memory, and about expertise in general: We don't remember isolated facts; we remember things in context. A board of randomly arranged chess pieces has no context—there are no similar boards to compare it to, no past games that it resembles, no ways to meaningfully chunk it. Even to the world's best chess player it is, in essence, noise.

In the same way that a few pages ago we used our knowledge of historic dates to chunk the twelve-digit number, chess masters use the vast library of chess patterns that they've cached away in long-term memory to chunk the board. At the root of the chess master's skill is that he or she simply has a richer vocabulary of chunks to recognize. Which is why it is so rare for anyone to achieve world-class status in chess—or any other field—without years of experience. Even Bobby Fischer, perhaps the greatest chess prodigy of all time, had been playing intensely for nine years before he was recognized as a grand master at age fifteen.

Contrary to all the old wisdom that chess is an intellectual activity based on analysis, many of the chess master's important decisions about which moves to make happen in the immediate act of perceiving the board. Like the chicken sexer who looks at the chick and simply sees its gender or the SWAT officer who immediately notices the bomb, the chess master looks at the board and simply sees the most promising move. The process usually happens within five seconds, and you can actually see it transpiring in the brain. Using magnetoencephalography, a technique that measures the weak magnetic fields given off by a thinking brain, researchers have found that higher-rated chess players are more likely to engage the frontal and parietal cortices of the brain when they look at the board, which suggests that they are recalling information from long-term memory. Lower-ranked players are more likely to engage the medial temporal lobes, which suggests that they are encoding new information. The experts are interpreting the present board in term of their massive knowledge of past ones. The lower-ranked players are seeing the board as something new.

Though chess might seem like a trivial subject for a psychologist to study—it is, after all, just a game—De Groot believed that his experiments with chess masters had much larger implications. He argued that expertise in "the field of shoemaking, painting, building, [or] confectionary" is the result of the same accumulation of "experiential linkings." According to Ericsson, what we call expertise is really just "vast amounts of knowledge, pattern-based retrieval, and planning mechanisms acquired over many years of experience in the associated domain." In other words, a great memory isn't just a by-product of expertise; it is the *essence* of expertise.

Whether we realize it or not, we are all like those chess masters and chicken sexers, interpreting the present in light of what we've learned in the past, and letting our previous experiences shape not only how we perceive our world, but also the moves we end up making in it.

Too often we talk about our memories as if they were banks into which we deposit new information when it comes in, and from which we withdraw old information when we need it. But that metaphor doesn't reflect the way our memories really work. Our memories are always with us, shaping and being shaped by the information flowing through our senses, in a continuous feedback loop. Everything we see, hear, and smell is inflected by all the things we've seen, heard, and smelled in the past.

In ways as obscure as sexing chickens and as profound as diagnosing an illness, who we are and what we do is fundamentally a function of what we remember. But if interpreting the world and acting in it are rolled up in the act of remembering, what about Ed and Lukas and other mental athletes I'd met? How did this supposedly "simple" technique called the memory palace grant them expert memories without their being experts in anything?

Even if Ericsson and his grad students wouldn't give me the results of all the tests I spent three days laboring on, I took enough notes on my performance to escape with some sense of where my baseline abilities stood. My digit span was about nine (above average, but nothing extraordinary), my ability to memorize poetry was abysmal, and I had not a clue when Confucius lived (though I did know what a carburetor was for). When I got back from Tallahassee, there was an e-mail waiting in my in-box from Ed:

> Hey there star-pupil, I know that you've been keeping training to a minimum until after the Florida people have put you through your paces. Very well done—that's

admirable in at least the sense that it will make for better science. But the next championships aren't a million miles away so you're going to have to begin preparing yourself pronto. Better get some pep from me now: You need to get your head towards the grindstone and enjoy leaving it there.

FOUR

THE MOST FORGETFUL MAN IN THE WORLD

Having met some of the best memories in the world, I decided that my next step would be to try to seek out the worst. What better way to try to begin to understand the nature and meaning of human memory than to investigate its absence? I went back to Google in search of Ben Pridmore's counterpart in the record books of forgetfulness, and dug up an article in *The Journal of Neuroscience* about an eighty-four-year-old retired lab technician called EP, whose memory extended back only as far as his most recent thought. He had one of the most severe cases of amnesia ever documented.

A few weeks after returning from Tallahassee, I phoned a neuroscientist and memory researcher named Larry Squire at the University of California, San Diego, and the San Diego VA Medical Center. Squire had been studying EP for over a decade, and agreed to bring me along on one of his visits to the bright bungalow in suburban San

Diego where EP lives with his wife. We traveled there with Jen Frascino, the research coordinator in Squire's lab who visits EP regularly to administer cognitive tests. Even though Frascino has been to EP's home some two hundred times, he greets her as a total stranger every time.

EP is six-foot-two, with perfectly parted white hair and unusually long ears. He's personable, friendly, gracious. He laughs a lot. He seems at first like your average genial grandfather. Frascino, a tall, athletic blonde, sits down with me and Squire opposite EP at his dining room table and asks a series of questions that are meant to gauge his basic knowledge and common sense. She quizzes him about what continent Brazil is on, the number of weeks in a year, the boiling temperature of water. She wants to demonstrate what a battery of cognitive tests has already proved: EP has a working knowledge of the world. His IQ is 103, and his short-term memory is entirely unimpaired. He patiently answers the questions—all correctly—with roughly the same sense of bemusement I imagine I would have if a total stranger walked into my house and earnestly asked me if I knew the boiling point of water.

"What is the thing to do if you find an envelope in the street that is sealed, addressed, and has a stamp on it?" Frascino asks.

"Well, you'd put it in the mailbox. What else?" He chuckles and shoots me a knowing, sidelong glance, as if to say, "Do these people think I'm an idiot?" But sensing that the situation calls for politeness, he turns back to Frascino and adds, "But that's a really interesting question you've got there. Really interesting." He has no idea he's heard it many times before.

"Why do we cook food?"

"Because it's raw?" The word raw carries his voice clear across the tonal register, his bemusement giving way to incredulity.

I ask EP if he knows the name of the last president.

"I'm afraid it's slipped my mind. How strange."

"Does the name Bill Clinton sound familiar?"

"Of course I know Clinton! He's an old friend of mine, a scientist, a good guy. I worked with him, you know."

He sees my eyes widen in disbelief and stops himself.

"Unless, that is, there's another Clinton around that you're thinking of—"

"Well, you know, the last president was named Bill Clinton also."

"He was? I'll be—!" He slaps his thigh and chuckles, but doesn't seem all that embarrassed.

"Who's the last president you remember?"

He takes a moment to search his brain. "Let's see. There was Franklin Roosevelt . . ."

"Ever heard of John F. Kennedy?"

"Kennedy? Hmm, I'm afraid I don't know him."

Frascino interjects with another question. "Why do we study history?"

"Well, we study history to know what happened in the past."

"But why do we want to know what happened in the past?"

"Because it's just interesting, frankly."

In November 1992, EP came down with what seemed like a mild case of the flu. For five days he lay in bed, feverish and lethargic, unsure of what was wrong, while inside his head a vicious virus known as herpes simplex was chewing its way through his brain, coring it like an apple. By the time the virus had run its course, two walnut-size chunks of brain matter in EP's medial temporal lobes had disappeared, and with them most of his memory.

The virus struck with freakish precision. The medial temporal lobes—there's one on each side of the brain—include the hippocampus and several adjacent regions that together perform the magical feat of turning our perceptions into long-term memories. Memories aren't actually stored in the hippocampus—they reside elsewhere, in the brain's corrugated outer layers, the neocortex—but the hippocampal area makes them stick. EP's hippocampus was destroyed, and without it he is like a camcorder without a working tape head. He sees, but he doesn't record.

EP has two types of amnesia—anterograde, which means he can't form new memories, and retrograde, which means he can't recall old memories either, at least not since about 1950. His childhood, his service in the merchant marine, World War II—all that is perfectly vivid. But as far as he knows, gas costs a quarter a gallon, and man never took that small step onto the moon.

Even though EP has been an amnesic for a decade and a half, and his condition has neither worsened nor improved, there's still much that Squire and his team hope to learn from him. A case like his, in which nature performs a cruel but perfect experiment, is, to put it crassly, a major boon to science. In a field in which so many basic questions are still unanswered, there is a limitless number of tests that can be performed on a mind like EP's. Indeed, there are only a handful of other individuals in the world in whom both hippocampi and the key adjacent structures have been so precisely notched out of an otherwise intact brain. Another severely amnesic case is Clive Wearing, a former music producer for the BBC who was struck by herpes encephalitis in 1985. Like EP's, his mind has become a sieve. Each time he greets his wife, it's as though he hasn't seen her in twenty years. He leaves her agonizing phone messages begging to be picked up from the nursing home where he lives. He also keeps an exhaustive diary that has become a tangible record of his daily anguish. But even the diary he finds hard to trust since—like every other object in his life—it is completely unfamiliar. Every time he opens it, it must feel like confronting a past life. It is filled with entries like this one:

~~8:31 AM: Now I am really, completely awake.~~
~~9:06 AM: Now I am perfectly, overwhelmingly awake.~~
9:34 AM: Now I am superlatively, actually awake.

Those scratched-out entries suggest an awareness of his condition that EP, perhaps blissfully, lacks. From across the table, Squire asks EP how his memory is doing these days.

"It's fair. Hard to say it's real good or bad."

EP wears a metal medical alert bracelet around his left wrist. Even though it's obvious what it's for, I ask him anyway. He turns his wrist over and casually reads it.

"Hmm. It says memory loss."

EP doesn't even remember that he has a memory problem. That is something he discovers anew every moment. And since he forgets that he always forgets, every lost thought seems like just a casual slip—an annoyance and nothing more—the same way it would to you or me.

"There's nothing wrong with him in his mind. That's a blessing," his wife, Beverly, tells me later, while EP sits on the couch, out of earshot. "I suppose he must know something is wrong, but it doesn't come out in conversation or in his way of life. But underneath he must know. He just must."

When I hear those words, I'm stung by the realization of how much more than just memories have been lost. Even EP's own wife can no longer access his most basic emotions and thoughts. Which is not to say that he doesn't have emotions or thoughts. Moment to moment, he certainly does. When informed of the births of his grandchildren, EP's eyes welled up each time—and then he promptly forgot that they existed. But without the ability to compare today's feelings to yesterday's, he cannot tell any cohesive narrative about himself, or about those around him, which makes him incapable of providing even the most basic psychological sustenance to his family and friends. After all, EP can only remain truly interested in anyone or anything for as long as he can maintain his attention. Any rogue thought that distracts him effectively resets conversation. A meaningful relationship between two people cannot sustain itself only in the present tense.

Ever since his sickness, space for EP has existed only as far as he can see it. His social universe is only as large as the people in the room. He lives under a narrow spotlight, surrounded by darkness. On a typical morning, EP wakes up, has breakfast, and returns to bed to listen to the radio. But back in bed, it's not always clear whether he's just had breakfast or just woken up. Often he'll have breakfast again, and return to bed to listen to some more radio. Some mornings he'll have breakfast for a third time. He watches TV, which can be very exciting from second to second, though shows with a clear beginning, middle, and end can pose a problem. He prefers the History Channel, or anything about World War II. He takes walks around the neighborhood, usually several times before lunch, and sometimes for as long as three quarters of an hour. He sits in the yard. He reads the newspaper, which must feel like stepping out of a time machine. Iraq? Internet? By the time EP gets to the end of a headline, he's usually forgotten how it began. Most of the time, after reading the weather, he just doodles on the paper, drawing mustaches on the photographs or tracing his spoon. When he sees home prices in the real estate section, he invariably announces his shock.

Without a memory, EP has fallen completely out of time. He has no stream of consciousness, just droplets that immediately evaporate. If you were to take the watch off his wrist—or, more cruelly, change the time—he'd be completely lost. Trapped in this limbo of an eternal present, between a past he can't remember and a future he can't contemplate, he lives a sedentary life, completely free from worry. "He's happy all the time. Very happy. I guess it's because he doesn't have any stress in his life," says his daughter Carol, who lives nearby. In his chronic forgetfulness, EP has achieved a kind of pathological enlightenment, a perverted vision of the Buddhist ideal of living entirely in the present.

"How old are you now?" Squire asks him.

"Let's see, fifty-nine or sixty. You got me," he says, raising his eyebrow contemplatively, as if he were making a calculation and not a guess. "My memory is not that perfect. It's pretty good, but sometimes people ask me questions that I just don't get. I'm sure you have that sometimes."

"Sure I do," says Squire kindly, even though EP's almost a quarter of a century off.

Without time, there would be no need for a memory. But without a memory, would there be such a thing as time? I don't mean time in the sense that, say, physicists speak of it: the fourth dimension, the independent variable, the quantity that dilates when you approach the speed of light. I mean psychological time, the tempo at which we experience life's passage. Time as a mental construct. Watching EP struggle to recount his own age, I recalled one of the stories Ed Cooke had told me about his research at the University of Paris when we met at the USA Memory Championship.

"I'm working on expanding subjective time so that it feels like I live longer," Ed had mumbled to me on the sidewalk outside the Con Ed headquarters, a cigarette dangling from his mouth. "The idea is to avoid that feeling you have when you get to the end of the year and feel like, where the hell did that go?"

"And how are you going to do that?" I asked.

"By remembering more. By providing my life with more chronological landmarks. By making myself more aware of time's passage."

I told him that his plan reminded me of Dunbar, the pilot in Joseph

Heller's *Catch-22* who reasons that since time flies when you're having fun, the surest way to slow life's passage is to make it as boring as possible.

Ed shrugged. "Quite the opposite. The more we pack our lives with memories, the slower time seems to fly."

Our subjective experience of time is highly variable. We all know that days can pass like weeks and months can feel like years, and that the opposite can be just as true: A month or year can zoom by in what feels like no time at all.

Our lives are structured by our memories of events. Event X happened just before the big Paris vacation. I was doing Y in the first summer after I learned to drive. Z happened the weekend after I landed my first job. We remember events by positioning them in time relative to other events. Just as we accumulate memories of facts by integrating them into a network, we accumulate life experiences by integrating them into a web of other chronological memories. The denser the web, the denser the experience of time.

It's a point well illustrated by Michel Siffre, a French chronobiologist (he studies the relationship between time and living organisms) who conducted one of the most extraordinary acts of self-experimentation in the history of science. In 1962, Siffre spent two months living in total isolation in a subterranean cave, without access to clock, calendar, or sun. Sleeping and eating only when his body told him to, he sought to discover how the natural rhythms of human life would be affected by living "beyond time."

Very quickly Siffre's memory deteriorated. In the dreary darkness, his days melded into one another and became one continuous, indistinguishable blob. Since there was nobody to talk to, and not much to do, there was nothing novel to impress itself upon his memory. There were no chronological landmarks by which he could measure the passage of time. At some point he stopped being able to remember what happened even the day before. His experience in isolation had turned him into EP. As time began to blur, he became effectively amnesic. Soon, his sleep patterns disintegrated. Some days he'd stay awake for thirty-six straight hours, other days for eight—without being able to tell the difference. When his support team on the surface finally called down to him on September 14, the day his experiment was scheduled to wrap up, it was only August 20 in his journal. He thought only a month had gone by. His experience of time's passage had compressed by a factor of two.

Monotony collapses time; novelty unfolds it. You can exercise daily and eat healthily and live a long life, while experiencing a short one. If you spend your life sitting in a cubicle and passing papers, one day is bound to blend unmemorably into the next—and disappear. That's why it's important to change routines regularly, and take vacations to exotic locales, and have as many new experiences as possible that can serve to anchor our memories. Creating new memories stretches out psychological time, and lengthens our perception of our lives.

William James first wrote about the curious warping and foreshortening of psychological time in his *Principles of Psychology* in 1890: "In youth we may have an absolutely new experience, subjective or objective, every hour of the day. Apprehension is vivid, retentiveness strong, and our recollections of that time, like those of a time spent in rapid and interesting travel, are of something intricate, multitudinous and long-drawn-out," he wrote. "But as each passing year converts some of this experience into automatic routine which we hardly note at all, the days and the weeks smooth themselves out in recollection to contentless units, and the years grow hollow and collapse." Life seems to speed up as we get older because life gets less memorable as we get older. "If to remember is to be human, then remembering more means being more human," said Ed.

There is perhaps a bit of Peter Pan to Ed's quest to make his life maximally memorable, but of all the things one could be obsessive about collecting, memories of one's own life don't seem like the most unreasonable. There's something even strangely rational about it. There's an old philosophical conundrum that often gets bandied about in introductory philosophy courses: In the nineteenth century, doctors began to wonder whether the general anesthetic they had been administering to patients might not actually put the patients to sleep so much as freeze their muscles and erase their memories of the surgery. If that were the case, could the doctors be said to have done anything wrong? Like the proverbial tree that falls without anyone hearing it, can an experience that isn't remembered be meaningfully said to have happened at all? Socrates thought the unexamined life was not worth living. How much more so the unremembered life?

Much of what science knows about memory was learned from a damaged brain remarkably similar to EP's. It belonged to another amnesic named Henry Molaison, who went by the initials HM and spent most of his life in a nursing home in Connecticut before dying in 2008. (Individuals in the medical literature always go by initials to protect their identities. HM's name was revealed after his death.) As a child, HM suffered from epilepsy, which began after a bike accident at age nine. By the time he was twenty-seven, he was blacking out several times a week and unable to do much of anything. A neurosurgeon named William Scoville thought he could relieve HM's symptoms with an experimental surgery that would excise the part of the brain that he suspected was causing the problem.

In 1953, while HM lay awake on the operating table, his scalp anesthetized, Scoville drilled a pair of holes just above the patient's eyes. The surgeon lifted the front of HM's brain with a small metal spatula while a metal straw sucked out most of the hippocampus, along with much of the surrounding medial temporal lobes. The surgery reduced the number of HM's seizures, but there was a tragic side effect: It soon became clear that he'd also been robbed of his memory.

Over the next five decades, HM was the subject of countless experiments and became the most studied patient in the history of brain science. Given the horrific outcome of Scoville's surgery, everyone assumed HM would be a singular case study.

EP shattered that assumption. What Scoville did to HM with a metal straw, nature did to EP with herpes simplex. Side by side, the grainy black-and-white MRIs of their brains are uncannily similar, though EP's damage is a bit more extensive. Even if you have no idea what a normal brain ought to look like, the two gaping symmetrical holes stare back at you like a pair of shadowy eyes.

Like EP, HM was able to hold on to memories just long enough to think about them, but once his brain moved on to something else, he could never bring them back. In one famous experiment conducted by the Canadian neuroscientist Brenda Milner, HM was asked to remember the number 584 for as long as possible. He spoke aloud as he was doing it:

> It's easy. You just remember 8. You see, 5, 8, and 4 add to 17. You remember 8, subtract it from 17 and it leaves 9. Divide 9 in half and you get 5 and 4 and there you are: 584. Easy.

He concentrated on this elaborate mantra for several minutes. But as soon as he was distracted, the number dissolved. He couldn't even remember that he'd been asked to remember something. Though scientists had known that there was a difference between long- and short-term memory since the late nineteenth century, they now had evidence in HM that the two types of memory processes happened in different parts of the brain, and that without most of the hippocampal area, HM couldn't turn a short-term memory into a long-term one.

Researchers also learned more about another kind of remembering from HM. Even though he couldn't say what he'd had for breakfast or name the current president, there were some things that he could recall. Milner found that he could learn complicated tasks without even realizing it. In one landmark study in 1962, she showed that HM could learn how to trace inside a five-pointed star on a piece of paper while looking at its reflection in a mirror. Each time Milner gave HM the task, he claimed never to have tried it before. And yet, each day his brain got better at guiding his hand to work in reverse. Despite his amnesia, he was remembering.

Subsequent studies of amnesics, including tests conducted on EP, have found that people who lose their memories are still capable of yet other kinds of unremembered learning. In one experiment, Squire gave EP a list of twenty-four words to memorize. As expected, within a few minutes, EP had no recollection of any of the words, or even that the exercise had happened at all. When asked whether he'd seen a given word before, he answered correctly only half the time. But then Squire sat EP in front of a computer monitor and gave him a different test. This time, forty-eight words were flashed on the screen for twenty-five milliseconds each, just long enough for the eye to catch some, but not all, of them (an eye blink, by comparison, happens in 100 to 150 milliseconds). Half the words were from the list that EP had read over and forgotten, and half were new. Squire asked EP to read each word after it flashed on the screen. Surprisingly, EP was far better at reading the words he'd seen before than the ones that were new. Even though he had no conscious recollection of them, somewhere in the recesses of his brain they had left an impression.

This phenomenon of unconscious remembering, known as priming, is evidence of an entire shadowy underworld of memories lurking beneath the surface of our conscious reckoning. Though there is disagreement about just how many memory systems there are, scientists generally divide memories broadly into two types: declarative and nondeclarative (sometimes referred to as explicit and implicit). Declarative memories are things you know you remember, like the color of your car, or what happened yesterday afternoon. EP and HM had lost the ability to make new declarative memories. Nondeclarative memories are the things you know unconsciously, like how to ride a bike or how to draw a shape while looking at it in a mirror (or what a word flashed rapidly across a computer screen means). Those unconscious memories don't seem to pass through the same short-term memory buffer as declarative memories, nor do they depend on the hippocampal region to be consolidated and stored. They rely primarily on different parts of the brain. Motor skill learning takes place largely in the cerebellum, perceptual learning in the neocortex, habit learning in the basal ganglia. As EP and HM have so strikingly demonstrated, you can damage one part of the brain, and the rest will keep on working. Indeed, most of who we are and how we think—the core material of our personalities—is bound up in implicit memories that are off-limits to the conscious brain.

Within the category of declarative memories, psychologists make a further distinction between semantic memories, or memories for facts and concepts, and episodic memories, or memories of the experiences of our own lives. Recalling that I had eggs for breakfast this morning would be an episodic memory. Knowing that breakfast is the first meal of the day is a semantic memory. Episodic memories are located in time and space: They have a where and a when attached to them. Semantic memories are located outside of time and space, as free-floating pieces of knowledge. These two different types of remembering seem to make use of different neural pathways, and rely on different regions of the brain, though both are critically dependent on the hippocampus and other structures within the medial temporal lobes. EP has lost both types of memory in equal measure, but curiously his forgetfulness extends back only for the last sixty or so years. His memories have faded along a gradient.

One of the many mysteries of memory is why an amnesic like EP should be able to remember when the atomic bomb fell on Hiroshima but not the much more recent fall of the Berlin Wall. For some unknown reason, it's the most recent memories that blur first in most amnesics, while distant memories retain their clarity. This phenomenon is known as Ribot's Law, after the nineteenth-century French psychologist who first noted it, and it's a pattern found also in Alzheimer's patients. It suggests something profound: that our memories are not static. Somehow, as memories age, their complexion changes. Each time we think about a memory, we integrate it more deeply into our web of other memories, and therefore make it more stable and less likely to be dislodged.

But in the process, we also transform the memory, and reshape it—sometimes to the point that our memories of events bear only a passing resemblance to what actually happened. Neuroscientists have only recently begun to observe this process happening inside the brain, but psychologists have understood for a long time that there are qualitative differences between old and new memories. Sigmund Freud first noted the curious fact that older memories are often remembered as if captured by a third person holding a camera, whereas more recent events tend to be remembered in the first person, as if through one's own eyes. It's as if things that happened to us become simply things that happened. Or as if, over time, the brain naturally turns episodes into facts.

How this process works at the level of neurons still remains a riddle. One well-supported hypothesis holds that our memories are nomadic. While the hippocampus is involved in their initial formation, their contents are ultimately held in long-term storage in the neocortex. Over time, as they are revisited and reinforced, memories are consolidated in a way that makes them impervious to erasure. They become entrenched in a network of cortical connections that allows them to exist independently of the hippocampus. All this raises a tantalizing question: Were EP's memories since 1950 completely obliterated when the virus ate its way through his medial temporal lobes, or did those memories just become inaccessible? Did the virus burn down half the house, or did it just throw away the key? We don't know.

It's thought that sleep plays a critical role in this process of consolidating our memories and drawing meaning out of them. Rats that have spent an hour running around a track apparently run through the same track in their sleep, and exhibit the same patterns of neural firings with their eyes closed as when they were learning the mazes in the first place. It has been suggested that the reason our own dreams so often feel like a surreal recombination of elements plucked from real life is that they are just the by-product of experiences slowly hardening into long-term memories.

Sitting with EP on the couch in his living room, I wonder if he still dreams. Of course, he can't tell, but I ask him anyway, just to see what he'll say. "From time to time," he tells me matter-of-factly, though his response is most certainly a confabulation. "But dreams are hard to remember."

We all come into the world as amnesics, and quite a few of us exit just the same. The other day, I was quizzing my three-year-old nephew about his second birthday party. Though the event took place more than a third of a lifetime ago, his recollections were surprisingly exact. He remembered the name of the young guitarist who had entertained him and his friends, and could recite some of the songs they had sung. He remembered the miniature drum set I'd given him as a gift. He remembered eating ice cream with cake. And yet, ten years from now, it is almost certain that he will remember none of this.

Until the age of three or four, almost nothing that happens to us leaves the sort of lasting impression that can be consciously recalled as an adult. The average age that people report having their earliest memory is three and a half, and those tend to be just blurry, fragmentary snapshots that are often false. How strange that during the period when a person is learning more rapidly than at any other point in his life—when one is learning to walk and talk and make sense of the world—so little of that learning is of the kind that is explicitly memorable.

Freud thought that infantile amnesia was a matter of adults repressing the hypersexualized fantasies of early childhood, which only become shameful in later life. I'm not sure you could find too many psychologists who still cling to that interpretation. The more likely explanation for this strange early forgetting lies in the fact that our brains are maturing rapidly during the first couple years of life, with unused neural connections getting pruned back, and new connections constantly forming. The neocortex is not fully developed until about the third or fourth year, around the time that children start laying down permanent memories. Anatomy, however, may only tell part of the story. As infants, we also lack schema for interpreting the world and relating the present to the past. Without experience—and perhaps most important, without the essential organizing tool of language—infants lack the capacity to embed their memories in a web of meaning that will make them accessible later in life. Those structures only develop over time, through exposure to the world. The vital learning that we do during the first years of life is virtually entirely of the implicit, nondeclarative kind. In other words, everyone on earth has had some taste of EP's condition. And like EP, we've all forgotten what it's like.

I'm curious to see EP's unconscious, nondeclarative memory at work, so I ask him if he's interested in taking me on a walk around his neighborhood. He says, "Not really," so I wait and ask him again a couple minutes later. This time he agrees. We walk out the front door into the high afternoon sun and turn right—his decision, not mine. I ask EP why we're not turning to the left instead.

"I'd just rather not go that way. This is just the way I go. I don't know why," he says.

If I asked him to draw a map of the route he takes at least three times a day, he'd never be able to do it. He doesn't even know his own address, or (almost as improbably for someone from San Diego) which way the ocean is. But after so many years of taking the same walk, the journey has etched itself on his unconscious. His wife, Beverly, now lets him go out alone, even though a single wrong turn would leave him completely lost. Sometimes he comes back from his walks with objects he's picked up along the way: a stack of round stones, a puppy, somebody's wallet. He can never explain how they came into his possession.

"Our neighbors love him because he'll come up to them and just start talking to them," Beverly tells me. Even though he thinks he's meeting them for the first time, he's learned through force of habit that these are people he should feel comfortable with, and he interprets those unconscious feelings of comfort as a good reason to stop and say hello.

That EP has learned to like his neighbors without ever learning who they are points to how many of our basic day-to-day actions are guided by implicit values and judgments, independent of declarative memory. I wonder what other things EP has learned through force of habit. What other nondeclarative memories have continued to shape him over the decade and a half since he lost his declarative memory? Surely, he must still have desires and fears, emotions, and cravings—even if his conscious recollection of those feelings is so fleeting that he cannot recognize them for long enough to verbalize them.

I thought of my own self fifteen years ago, and how much I've changed in the same period. The me who exists today and the me who existed then, if put side by side, would look more than vaguely similar. But we are a completely different collection of molecules, with different hairlines and waistlines, and, it sometimes seems, little in common besides our names. What binds that me to this me, and allows me to maintain the illusion that there is continuity from moment to moment and year to year, is some relatively stable but gradually evolving thing at the nucleus of my being. Call it a soul, or a self, or an emergent by-product of a neural network, but whatever you want to call it, that element of continuity is entirely dependent on memory.

But even if we are at the mercy of our memories in establishing our identities, it is clear that EP is much more than just a soulless golem. In spite of everything he's lost, there is still a person there, and a personality—a charming personality, in fact—with a unique perspective on the world. Even if a virus wiped clean his memories, it didn't completely wipe clean his personhood. It just left a hollow, static self that can never grow and can never change.

We cross the street and walk away from Beverly and Carol, leaving me alone with EP for the first time. He doesn't know who I am, or what I'm doing at his side, although he seems to sense that I'm there for some good reason. He looks at me and purses his lips, and I can see that he's searching for something to say. Rather than try to fill the empty silence, I let it linger for a moment to see where the discomfort might lead. I guess I'm hoping for some fleeting recognition of how odd it all must be, this scene without a prologue. But no such recognition comes, or if it does, EP never lets it surface. He is trapped, I realize, in the ultimate existential nightmare, utterly blind to the reality in which he lives. The impulse strikes me to help him escape, at least for a second. I want to take him by the arm and shake him. "You have a rare and debilitating memory disorder," I want to tell him. "The last fifty years have been lost to you. In less than a minute, you're going to forget that this conversation ever even happened." I imagine the horror that would descend upon him, the momentary clarity, the gaping emptiness that would open up in front of him, and close just as quickly. And then the passing car or the singing bird that would snap him back into his oblivious bubble. But of course I don't do it.

"We've gone far enough," I tell him, and point him in the direction from which we came. We turn around and walk back down the street whose name he's forgotten, past the waving neighbors he doesn't recognize, to a home he doesn't know. In front of the house sits a car with tinted windows. We turn to look at our reflections. I ask EP what he sees.

"An old man," he says. "That's all."

FIVE

THE MEMORY PALACE

I had arranged to get together with Ed one last time before he headed back to Europe. He wanted to meet me in Central Park, which he had never seen before, and which he insisted was a vital stop on his tour of America. After taking in the bare late-winter trees and watching the runners do their midday laps around the Reservoir, we ended up at the southern end of the park, directly across the street from the Ritz-Carlton Hotel. It was a frigid and brutally windy afternoon—less than ideal conditions for thinking of any kind, much less memorizing. Nevertheless, Ed insisted that we remain outdoors. He handed me his cane and gamely clambered up one of the big boulders near the edge of the park, with what appeared to be some pain in his chronically arthritic joints. After scanning the horizon and commenting on the "perfect sublimity" of the spot, he invited me to join him on top of the rock. He had promised that he could teach me a few basic memory techniques in under an hour. It was hard to imagine we could brave the weather for any longer than that.

"I have to warn you," Ed said, as he delicately seated himself cross-legged, "you are shortly going to go from having an awed respect for people with a good memory to saying, 'Oh, it's all a stupid trick.'" He paused and cocked his head, as if to see if that would in fact be my response. "And you will be wrong. It's an unfortunate phase you're just going to have to pass through."

He started his lesson with the most basic principle of all mnemonics: "elaborative encoding." Our memories weren't built for the modern world, he explained. Like our vision, our capacity for language, our ability to walk upright, and every other one of our biological faculties, our memories evolved through a process of natural selection in an environment that was quite different from the one we live in today.

Most of the evolution that shaped the primitive brains of our prehuman ancestors into the linguistic, symbolic, neurotic modern brains that serve us (sometimes poorly) today took place during the Pleistocene, an epoch which began about 1.8 million years ago and only ended ten thousand years ago. During that period—and in a few isolated places, still to this day—our species made its living as hunter-gatherers, and it was the demands of that lifestyle that sculpted the minds we have today.

Much as our taste for sugar and fat may have served us well in a world of scarce nutrition, but is now maladaptive in a world of ubiquitous fast food joints, our memories aren't perfectly adapted for our contemporary information age. The tasks that we often rely on our memories for today simply weren't relevant in the environment in which the human brain evolved. Our ancestors didn't need to recall phone numbers, or word-for-word instructions from their bosses, or the Advanced Placement USA history curriculum, or (because they lived in relatively small, stable groups) the names of dozens of strangers at a cocktail party.

What our early human and hominid ancestors *did* need to remember was where to find food and resources, and the route home, and which plants were edible and which were poisonous. Those are the sorts of vital memory skills that they depended on every day, and it was—at least in part—in order to meet those demands that human memory evolved as it did.

The principle underlying all memory techniques is that our brains don't remember all types of information equally well. As exceptional as we are at remembering visual imagery (think of the two-picture recognition test), we're terrible at remembering other kinds of information, like lists of words or numbers. The point of memory techniques is to do what the synasthete S did instinctually: to take the kinds of memories our brains aren't good at holding on to and transform them into the kinds of memories our brains were built for.

"The general idea with most memory techniques is to change whatever boring thing is being inputted into your memory into something that is so colorful, so exciting, and so different from anything you've seen before that you can't possibly forget it," Ed explained to me between breaths into his clenched fists. "That's what elaborative encoding is. In a moment, we're going to do this with a list of words, which is just a sort of general exercise for getting ahold of the techniques. Then you're going to be able to move on to numbers, playing cards, and then, from there, to complex concepts. Basically, when we're done with you, you're going to be able to learn anything you want to, really."

Ed recounted how on a recent visit to Vienna, he and Lukas had partied until dawn the night before Lukas's biggest exam of the year, and only stumbled home just before sunrise. "Lukas woke up at noon, learned everything for the exam in a memory blitz, and then passed it," said Ed. "When you're that effective at learning, it's a bit of a temptation to not bother oneself with feelings of academic guilt until the last possible moment. Lukas has figured out that effort is a rather vulgar exercise."

Ed tucked his curls behind his ears, and asked me what I wanted to memorize first. "We could start by learning something useful, like the Egyptian pharaohs or the terms of the American presidents," he offered. "Or perhaps a Romantic poem? We could do the geological epochs, if you'd like."

I laughed. "That all sounds *very* useful."

"We could quickly learn all the American football winners for the last century or so, or the point averages of the top baseball stars, if you'd like."

"Do you know—*really know*—all the winners of the Super Bowl?" I asked.

"Well, no, I don't. I prefer cricket. But I'd be happy to teach them to you. That's the point: We can quickly learn anything with these techniques. Look, you tempted or not?"

"I'm tempted."

"Well, I suppose the most obvious, practical use of this technique is the mastery of one's to-do list. Do you keep a to-do list?"

"At home, yes. Sort of. From time to time."

"I see. Well, I keep a to-do list in my memory at all times. We'll use mine."

Ed asked for a piece of paper, which he then scribbled a few words on. He handed it back to me with a mischievous smirk. It was a list of fifteen items. "Just a few things I've got to get done around town before I head upstate for a party a friend of mine is throwing," he said.

I read the list aloud:

Pickled garlic
Cottage cheese
Salmon (peat-smoked if poss.)

Six bottles of white wine
Socks (x3)
Three hula-hoops (spare?)
Snorkel
Dry ice machine
E-mail Sophia
Skin-toned cat suit
Find Paul Newman film—Somebody Up There Likes Me
Elk sausages??
Megaphone and director's chair
Harness and ropes
Barometer

"This list is from your memory?" I asked incredulously.

"From my memory it came. Into your memory it shall go," said Ed.

"And this is serious?"

"Well, I'm not sure if I'll be able to find everything on it. Do you have cottage cheese in New York?"

"I'm a little more concerned about the elk sausages and the skin-toned cat suit," I told him. "And besides, aren't you leaving town to go back to England tomorrow?"

"Yes, well, I'm prepared to accept that many of these items aren't *absolutely* necessary." He winked. "The point of this exercise, however, is that you are going to commit this list to memory."

Ed told me that by learning the techniques he was about to teach, I would be installing myself in a "proud tradition of mnemonists." That proud tradition began, at least according to legend, in the fifth century B.C. with the poet Simonides of Ceos standing in the rubble of the great banquet hall collapse in Thessaly. As the poet closed his eyes and reconstructed the crumbled building in his imagination, he had an extraordinary realization: He remembered where each of the guests at the ill-fated dinner had been sitting. Even though he had made no conscious effort to memorize the layout of the room, it had nevertheless left a durable impression upon his memory. From that simple observation, Simonides reputedly invented a technique that would form the basis of what came to be known as the art of memory. He realized that if it hadn't been guests sitting at the banquet table, but rather something else—say, every great Greek dramatist seated in order of their dates of birth—he would have remembered that instead. Or what if, instead of banquet guests, he saw each of the words of one of his poems arrayed around the table? Or every task he needed to accomplish that day? Just about anything that could be imagined, he reckoned, could be imprinted upon one's memory, and kept in good order, simply by engaging one's spatial memory in the act of remembering. To use Simonides' technique, all one has to do is convert something unmemorable, like a string of numbers or a deck of cards or a shopping list or *Paradise Lost*, into a series of engrossing visual images and mentally arrange them within an imagined space, and suddenly those forgettable items become unforgettable.

Virtually all the nitty-gritty details we have about classical memory training—indeed, nearly all the memory tricks in the mental athlete's arsenal—were first described in a short, anonymously authored Latin rhetoric textbook called the *Rhetorica ad Herennium*, written sometime between 86 and 82 B.C. It is the only truly complete discussion of the memory techniques invented by Simonides to have survived into the Middle Ages. Though the intervening two thousand years have seen quite a few innovations in the art of memory, the basic techniques have remained fundamentally unchanged from those described in the *Ad Herennium*. "This book is our bible," Ed told me.

Ed reads both Latin and ancient Greek (as well as speaking French and German fluently) and fancies himself an amateur classicist. The *Ad Herennium* was to be the first of several ancient texts he pressed upon me. Before I sampled Tony Buzan's expansive oeuvre (he's authored or coauthored over 120 books) or any of the self-help books put out by the top mental athletes, Ed wanted me to start my investigation with the classics. In addition to the *Ad Herennium*, there would be translated excerpts of Quintilian's *Institutio Oratoria* and Cicero's *De Oratore* for me to read, followed by a collection of medieval writings on memory by Thomas Aquinas, Albertus Magnus, Hugh of St. Victor, and Peter of Ravenna.

The techniques introduced in the *Ad Herennium* were widely practiced in the ancient world. In fact, in his own writings on the art of memory, Cicero says that the techniques are so well known that he felt he didn't need to waste ink describing them in detail (hence our reliance on the *Ad Herennium*). Once upon a time, every literate person was versed in the techniques Ed was about to teach me. Memory training was considered a centerpiece of classical education in the language arts, on par with grammar, logic, and rhetoric. Students were taught not just what to remember, but how to remember it.

In a world with few books, memory was sacrosanct. Just look at Pliny the Elder's *Natural History*, the first-century encyclopedia that chronicled all things wondrous and useful for winning bar bets in the classical world, including the most exceptional memories then known to history. "King Cyrus could give the names of all the soldiers in his army," Pliny reports. "Lucius Scipio knew the names of the whole Roman people. King Pyrrhus's envoy Cineas knew those of the Senate and knighthood at Rome the day after his arrival . . . A person in Greece named Charmadas recited the contents of any volumes in libraries that anyone asked him to quote, just as if he were reading them." There are plenty of reasons not to take everything Pliny says at face value (he also reported the existence of a race of dog-headed people in India) but the sheer volume of anecdotes about extraordinary memories in the classical world is itself telling. Seneca the Elder could repeat two thousand names in the order they'd been given to him. St. Augustine tells of a friend, Simplicius, who could recite Virgil by heart—backward. (That he could recite it forward seems to have been unremarkable.) A strong memory was seen as the greatest virtue since it represented the internalization of a universe of external knowledge. "Ancient and medieval people reserved their awe for memory. Their greatest geniuses they describe as people of superior memories," writes Mary Carruthers, the author of two books on the history of memory techniques. Indeed, the single most common theme in the lives of the saints—besides their superhuman goodness—is their often extraordinary memories.

The *Ad Herennium*'s discussion of memory—"that treasure-house of inventions and the custodian of all parts of rhetoric"—is actually quite short, about ten pages embedded in a far longer treatise on rhetoric and oration. It begins by making a distinction between natural memory and artificial memory: "The natural memory is that memory which is embedded in our minds, born simultaneously with thought. The artificial memory is that memory which is strengthened by a kind of training and system of discipline." In other words, natural memory is the hardware you're born with. Artificial memory is the software you run on your hardware.

Artificial memory, the anonymous author continues, has two basic components: images and places. Images represent the contents of what one wishes to remember. Places—or loci, as they're called in the original Latin—are where those images are stored.

The idea is to create a space in the mind's eye, a place that you know well and can easily visualize, and then populate that imagined place with images representing whatever you want to remember. Known as the "method of loci" by the Romans, such a building would later come to be called a "memory palace."

Memory palaces don't necessarily have to be palatial—or even buildings. They can be routes through a town—as they were for S—or station stops along a railway, or signs of the zodiac, or even mythical creatures. They can be big or small, indoors or outdoors, real or imaginary, so long as there's some semblance of order that links one locus to the next, and so long as they are intimately familiar. The four-time USA memory champion Scott Hagwood uses luxury homes featured in *Architectural Digest* to store his memories. Dr. Yip Swee Chooi, the effervescent Malaysian memory champ, used his own body parts as loci to help him memorize the entire 56,000-word, 1,774-page Oxford Chinese-English dictionary. One might have dozens, hundreds, perhaps even thousands of memory palaces, each built to hold a different set of memories.

In Australia and the American Southwest, Aborigines and Apache Indians independently invented forms of the loci method. But instead of using buildings, they relied on the local topography to plot their narratives, and sang them across the landscape. Each hillock, boulder, and stream held a part of the story. "Myth and map became coincident," says John Foley, a linguistic anthropologist at the University of Missouri who studies memory and oral traditions. One of the tragic consequences of embedding narrative into the landscape is that when Native Americans had land taken from them by the USA government, they lost not only their home but their mythology as well.

"The thing to understand, Josh, is that humans are very, very good at learning spaces," Ed remarked from his perch on the boulder. "Just to give an example, if you are left alone for five minutes in someone else's house you've never visited before, and you're feeling energetic and nosy, think about how much of that house could be fixed in your memory in that brief period. You'd be able to learn not just where all the different rooms are and how they connect with each other, but their dimensions and decoration, the arrangement of their contents, and where the windows are. Without really noticing it, you'd remember the whereabouts of hundreds of objects and all sorts of dimensions that you wouldn't even notice yourself noticing. If you actually add up all that information, it's like the equivalent of a short novel. But we don't ever register that as being a memory achievement. Humans just gobble up spatial information."

The principle of the memory palace, he continued, is to use one's exquisite spatial memory to structure and store information whose order comes less naturally—in this case, Ed's to-do list. "What you're going to find is that in the same way as it's impossible to get confused about the order of rooms in that house, it will be equally obvious that immediately after I locate three hula hoops, a snorkel, and a dry ice machine, my next task will be e-mailing my friend Sophia."

The crucial thing was to choose a memory palace with which I was intimately familiar. "For your first memory palace, I'd like you to use the house you grew up in, since that's a space you're likely to know very well," Ed said. "We're going to array the items of my to-do list one by one along a route that will snake around your childhood home. When it comes time to recall the list, all you will need to do is retrace the steps we're about to take in your imagination. The hope is that all the objects you're about to memorize will pop back into mind. Now, tell me, is your childhood home a bungalow?"

"More of a two-story brick house," I said.

"Does it have a cute postbox at the end of the driveway?"

"No. Why?"

"Shame. That would be an excellent first locus at which to deposit an image of the first item on our to-do list. But that's okay. We can start at the foot of the driveway. I want you to close your eyes and try to visualize in as much detail as possible a large bottle of pickled garlic standing right where the car should be parked."

I wasn't entirely sure what I was supposed to be visualizing. "What is pickled garlic? Is that, like, an English delicacy?" I asked.

"Um, no, it's just the sort of snack one brings along for a weekend out in the mountains." He flashed another impish grin. "Now, it's very important to try to remember this image multisensorily." The more associative hooks a new piece of information has, the more securely it gets embedded into the network of things you already know, and the more likely it is to remain in memory. Just as S spontaneously and involuntarily turned every sound that passed through his ears into a chorus of colors and smells, the author of the *Ad Herennium* urged his readers to do the same with every image they wanted to remember.

"It's important that you deeply process that image, so you give it as much attention as possible," Ed continued. "Things that grab our attention are more memorable, and attention is not something you can simply will. It has to be pulled in by the details. By laying down elaborate, engaging, vivid images in your mind, it more or less guarantees that your brain is going to end up storing a robust, dependable memory. So try to imagine the pleasant smell of the pickled garlic, and exaggerate its proportions. Imagine tasting it. Really let the flavor roll around on your tongue. And make sure you see yourself doing this at the foot of your driveway." If I didn't know what pickled garlic was, I was even less sure of how it tasted. Nevertheless, I imagined a large bottle of the stuff standing proudly at the foot of my parents' driveway.

(I'd encourage you, reader, to do the same along with me. Try imagining a bottle of pickled garlic at the foot of your own driveway, or if you don't have a driveway, someplace else outside your home. Really try to visualize it.)

"Now that you've installed a complete multisensorial picture of pickled garlic, we're going to walk up the path to your home and visualize the next item on our to-do list at the front door. It's cottage cheese. I want you to close your eyes and see an enormous wading-pool-size tub of cottage cheese. Have you got it?

"I think so."

(*Have you?)*

"Now I want you to imagine Claudia Schiffer swimming in this tub of cottage cheese. I want you to imagine her swimming in the buff, and dripping with dairy. Are you picturing this? I don't want you to miss any of the details here."

The *Ad Herennium* advises readers at length about creating the images for one's memory palace: the funnier, lewder, and more bizarre, the better. "When we see in everyday life things that are petty, ordinary, and banal, we generally fail to remember them, because the mind is not being stirred by anything novel or marvelous. But if we see or hear something exceptionally base, dishonorable, extraordinary, great, unbelievable, or laughable, *that* we are likely to remember for a long time."

The more vivid the image, the more likely it is to cleave to its locus. What distinguishes a great mnemonist, I was learning, is the ability to create these sorts of lavish images on the fly, to paint in the mind a scene so unlike any that has been seen before that it cannot be forgotten. And to do it quickly. Which is why Tony Buzan tells anyone who will listen that the World Memory Championship is less a test of memory than of creativity.

When forming images, it helps to have a dirty mind. Evolution has programmed our brains to find two things particularly interesting, and therefore memorable: jokes and sex—and especially, it seems, jokes about sex. (Do you remember what Rhea Perlman and Manute Bol were doing on the first page of this book?) Even memory treatises from comparatively prudish eras make this point. Peter of Ravenna, author of the most famous memory textbook of the fifteenth century, first asks the pardon of chaste and religious men before revealing "a secret which I have (through modesty) long remained silent about: if you wish to remember quickly, dispose the images of the most beautiful virgins into memory places; the memory is marvelously excited by images of women."

I was finding it a little hard to get excited about Claudia Schiffer and her tub of cottage cheese, however. My nose and ears were stinging from the icy wind. "Um, Ed, should we maybe take this lesson inside somewhere?" I asked. "There must be a Starbucks around here."

"No, no. This cold air is good for the brain," he said. "Now pay attention. We've just walked inside the door of your house. I want you to turn to the left in your mind's eye. What's the next room you enter into?" he asked.

"The living room. There's a piano in it."

"Perfect. Our third item is peat-smoked salmon. So let's imagine that underneath the strings of this piano there's a lot of smoking peat. And lying on top of the piano strings, there is a Hebridean salmon. Ooooh . . . can you smell that?" He whiffed at the cold air.

Again, I wasn't certain what peat-smoked salmon was, but it sounded like lox, so that's what I visualized. "Smells great," I said, my eyes still closed.

(If you don't have a piano in your own home, just put the peat-smoked salmon somewhere to the left of your front door.)

The next item on the list was six bottles of white wine, which we decided to place on the stained white couch next to the piano.

"Now, anthropomorphizing the bottles of wine is quite a good idea," Ed suggested. "Animate images tend to be more memorable than inanimate images." That advice, too, came from the *Ad Herennium*. The author instructs his readers to create images of "exceptional beauty or singular ugliness," to put them into motion, and to ornament them in ways that render them more distinct. One could "disfigure them, as by introducing one stained with blood or soiled with mud or smeared with red paint," or else proceed by "assigning certain comic effects to our images."

"Perhaps you should imagine the wines discussing their relative merits among themselves," Ed suggested.

"So, like, Mr. Merlot is saying—"

"Merlot is *not* a white wine, Josh," he interrupted, with a bemused titter. "Rather, let's imagine that the chardonnay is plaintively insulting the soil quality of the sauvignon blanc, while the gewürztraminer is giggling away nearby at the expense of the rieslings . . . That sort of thing."

I thought that was a funny image, and one sure to stick in my mind. But why? What makes six snooty, anthropomorphized wine bottles more memorable than the words "six bottles of wine"? Well, for one thing, visualizing such an outlandish image demanded more mental indulgence than simply reading four words. In the process of expending all that mental effort, I was forming more durable connections among the neurons that would encode that memory. But even more important, the memorableness of those talking wine bottles is a function of their novelty. While I have seen many wine bottles in my day, I have never seen one that talks before. Were I to simply try to remember the words "six bottles of wine," that memory would soon blend in with all my other memories of wine bottles.

Consider: How many of the lunches that you ate over the last week can you recall? Do you remember what you ate today? I hope so. Yesterday? I bet it takes a moment's effort. And what about the day before yesterday? What about a week ago? It's not so much that your memory of last week's lunch has disappeared; if provided with the right cue, like where you ate it, or whom you ate it with, you would likely recall what had been on your plate. Rather, it's difficult to remember last week's lunch because your brain has filed it away with all the other lunches you've ever eaten as *just another lunch*. When we try to recall something from a category that includes as many instances as "lunch" or "wine," many memories compete for our attention. The memory of last Wednesday's lunch isn't necessarily gone; it's that you lack the right hook to pull it out of a sea of lunchtime memories. But a wine that talks: That's unique. It's a memory without rivals.

"Next along on our list we have three pairs of socks," Ed continued. "Maybe there's a lamp you can hang them on nearby?"

"Yeah, there's a lamp right next to the couch," I said.

(If you're still following along, you should be putting those six bottles of wine and three pairs of socks somewhere in the first room of your house.)

"Splendid. Now, I know of precisely two ways to make socks attention-grabbing. One is to have them be appallingly old and smelly. The other is to make them those incredible socks made of cotton in nice colors that you can never really find. Let's make these socks the latter. So I'd like you to just see them dangling there on the lamp. And since it's often good to have a bit of supernatural crap going on, too, perhaps you can imagine that there is an elegant ghost inside the socks that is stretching and pulling them. Really try to see it. Imagine the feeling of those soft cotton socks coolly brushing against your forehead."

I followed Ed like this around my childhood home, dropping images along the way as I sauntered from room to room in my imagination. In the dining room, I visualized three hula-hooping women on top of the table. Stepping into the kitchen, I saw a man wearing a snorkel diving into the sink, and a dry-ice machine blowing smoke across the counter. *(Are you keeping up?)* From there, I moved into the den. The next item on the list was "e-mail Sophia."

I unclenched my eyes to ask Ed for help, and watched him licking the edge of a rolling paper for a fresh cigarette. "What should 'e-mail Sophia' look like?" I asked.

"Ooh, that's a tough one," he said, putting down his cigarette. "You see, e-mail isn't very memorable in itself. The more abstract the word, the less memorable it is. We need to make e-mail concrete somehow." Ed paused and thought on it for a moment. "What I'd like to propose is that you imagine a she-male sending the e-mail. Can you do that? And you'll need to associate that she-male with Sophia. What's the first image that enters your mind when I say the word 'Sophia'?"

"It's the capital of Bulgaria," I said.

"That's very educated of you, Josh. Bravo. But, alas, not very memorable. Instead let's make it Sophia Loren. And let's have her sitting on the lap of the she-male as she/he types away at the computer. Have you visualized it? Are you sufficiently engaged by this image? Splendid."

The pace of image-making now picked up. I left the den and visualized a comely woman in a skin-toned cat suit purring in the hallway. I placed Paul Newman in a nearby alcove, and an elk at the top of the stairs to the basement. I walked down the stairs and into the garage, where I left behind an image of Ed sitting in a director's chair barking orders through a giant megaphone. Then I imagined myself pressing the clicker that raises the garage door and walking out into the backyard, where a harnessed climber was using ropes to ascend a sizable oak tree. And the final image, a barometer, was installed alongside the backyard fence. "To remind you that it's a BAR-ometer, you should see a thermometerlike column sitting in a bed of pork scratchings and other bar snacks," Ed helpfully suggested. Having completed my circuit of the house, I opened my eyes.

"Well done," Ed said, with slow and deliberate applause. "Now, I think you're going to find that the process of recalling these memories is incredibly intuitive. See, normally memories are stored more or less at random in semantic networks, or webs of association. But you have now stored a large number of memories in a very controlled context. Because of the way spatial cognition works, all you have to do is retrace your steps through your memory palace, and hopefully at each point the images you laid down will pop back into your mind as you pass by them. All you'll have to do is translate those images back into the things you were trying to learn in the first place."

I closed my eyes again and saw myself back at the foot of my parents' driveway. The enormous jar of pickled garlic was just where I'd left it. I walked up the path to the front door. There was Claudia

Schiffer, seductively scrubbing herself with a sponge in a tub of cottage cheese. I opened the door and turned to the left, and inhaled a noseful of the fish that was still laid out across the strings of the piano, curing in peat smoke. I felt its flavor on my tongue. I could hear the high-pitched chatter of those haughty wine bottles on the couch, and feel the three pairs of luxurious cotton socks on the lamp brushing softly against my forehead. I couldn't believe it was really working. I called out the first five items of the to-do list for Ed to confirm. "Pickled garlic! Cottage cheese! Peat-smoked salmon! Six bottles of wine! Three pairs of socks!"

"Exceptional!" Ed shouted into the cold wind. "Exceptional! The makings of KL7 material here!"

Well, I knew my performance couldn't have been *that* exceptional, given the much more impressive feats I'd witnessed the day before. Still, I was feeling pretty good about my accomplishment. I continued walking through the house, picking up the bread crumbs of exotic images I'd deposited earlier. "Three hula hoops on the dining room table! Snorkel in the sink! Dry ice machine on the counter!" To my surprise and delight, all fifteen images were exactly where I'd left them. But would those memories really stick, I wondered? A week from now, would I still remember Ed's to-do list?

"Barring an episode of binge drinking or a wallop to the side of your head, you're going to find that those images will hold in your mind far longer than you might expect," Ed promised me. "And if you revisit the journey through your memory palace later this evening, and again tomorrow afternoon, and perhaps again a week from now, this list will leave a truly lasting impression. And having now done this with fifteen words, we could easily do it with fifteen hundred, provided you had an appropriately large memory palace to store them in. And then having mastered random words, we can move onto the truly fun stuff, like playing cards and Heidegger's *Being and Time*."

SIX

HOW TO MEMORIZE A POEM

My first assignment was to begin collecting architecture. Before I could embark on any serious degree of memory training, I first needed a stockpile of memory palaces at my disposal. I went for walks around the neighborhood. I visited friends' houses, the local playground, Oriole Park at Camden Yards in Baltimore, the East Wing of the National Gallery of Art. And I traveled back in time: to my high school, to my elementary school, to the house on Reno Road where my family lived until I was four years old. I focused on wallpaper and the arrangement of furniture. I tried to feel the flooring under my feet. I reminded myself of emotionally resonant incidents that occurred in each room. And then I carved each building up into loci that would serve as cubbyholes for my memories. The goal, as Ed explained it, was to know these buildings so thoroughly—to have such a rich and textured set of associations with every corner of every room—that when it came time to learn some new body of information, I could speed through my palaces, scattering images as quickly as I could sketch them in my imagination. The better I knew the buildings, and the more each felt like home, the stickier my images would be, and the easier it would be to reconstruct them later. Ed figured I'd need about a dozen memory palaces just to begin my training. He has several hundred, a metropolis of mental storehouses.

At this point, out of full disclosure, I ought to say a word or two about my living arrangements at the time that I began my dalliance with memory training. I was a recent college grad trying to make it as a journalist, sponging off my parents in the home in Washington, D.C., where I'd grown up. I was sleeping in my childhood bedroom with a pair of Baltimore Orioles pennants above the window and a book of Shel Silverstein's poems on the shelf, and working in a makeshift office in the basement, at a desk I'd set up between my father's Nordic Track and a stack of boxes filled with old family photos.

My office was awash in Post-it notes, and long lists of items I needed to catch up on: calls to be returned, article ideas to be investigated, personal and professional chores to be completed. Fortified with confidence from my successes in Central Park, I tore down a handful of the most urgent items, converted them into images, and diligently filed them away in a memory palace I had constructed out of my grandmother's suburban ranch home. "Get car inspected" became an image of Inspector Gadget circling the old Buick in her driveway. "Find book on African kings" was an occasion to imagine Shaka Zulu hurling a spear at her front door. "Book Phoenix ticket" led me to transform her living room into a landscape of desert and canyons, and to picture a phoenix rising from the ashes of her antique credenza. This was all well and good, and even kind of fun, but it was also exhausting. I noticed, upon memorizing ten or so of my Post-it notes, that I felt physically tired, like my mind's eye was getting bloodshot.

This was harder work than it seemed, and much less efficient than I'd imagined. And there were still a few items on the wall I had no clue what to do with. How was I supposed to turn telephone numbers into images? What was I supposed to do with e-mail addresses? I fell back into my office chair with a handful of Post-its clinging to my palm and looked up at my wall, whose off-white paint now showed through in a few additional patches, and wondered what, really, was the point of all this. In truth, those notes had been working just fine stuck to my wall. Surely the art of memory had more valuable applications.

I stood up and pulled a copy of the *Norton Anthology of Modern Poetry* off my bookshelf. It was an 1,800-page brick of a book that I had purchased once upon a time at a used bookstore and had opened not more than twice since. If the ancient art of memory was good for anything, I figured, surely it was learning poetry by heart. Simonides, I knew, was not a hero of the ancient world for having discovered a clever way to remember his to-do lists. His discovery was meant to serve a humanizing agenda. And what could be more humanizing than committing poetry to memory?

Ed, I had already discovered, was always memorizing something. He had long ago learned the bulk of *Paradise Lost* by heart (at the rate of two hundred lines per hour, he told me), and had been slowly slogging his way through Shakespeare. "My philosophy of life is that a heroic person should be able to withstand about ten years in solitary confinement without getting terribly annoyed," he said. "Given that an hour of memorization yields about ten solid minutes of spoken poetry, and those ten minutes have enough content to keep you busy for a full day, I figure you can squeeze at least a day's fun out of each hour of memorization—if you should ever happen to find yourself in solitary confinement."

This worldview owes a lot to the collection of ancient and medieval texts on memory that Ed had relentlessly tried to foist upon me. For those early writers, a trained memory wasn't just about gaining easy access to information; it was about strengthening one's personal ethics and becoming a more complete person. A trained memory was the key to cultivating "judgment, citizenship, and piety." What one memorized helped shape one's character. Just as the secret to becoming a chess grand master was to learn old games, the secret to becoming a grand master of life was to learn old texts. In a tight spot, where could one look for guidance about how to act, if not the depths of memory? Mere reading is not necessarily learning—a fact that I am personally confronted with every time I try to remember the contents of a book I've just put down. To *really* learn a text, one had to memorize it. As the early-eighteenth-century Dutch poet Jan Luyken put it, "One book, printed in the Heart's own wax / Is worth a thousand in the stacks."

The ancient and medieval way of reading was very different from how we read today. One didn't just memorize texts; one ruminated on them—chewed them up and regurgitated them like cud—and in the process, became intimate with them in a way that made them one's own. As Petrarch said in a letter to a friend, "I ate in the morning what I would digest in the evening; I swallowed as a boy what I would ruminate upon as an older man. I have thoroughly absorbed these writings, implanting them not only in my memory but in my marrow." Augustine was said to be so steeped in the Psalms that they, as much as Latin itself, comprised the principle language in which he wrote.

This was an attractive fantasy: I imagined that if I could only learn to memorize like Simonides, I would be able to commit reams of poetry to heart. I could make a clean sweep through the best verse and *really* absorb it. I imagined becoming one of those admirable (if sometimes insufferable) individuals who always seem to have an apposite quotation to drop into conversation. I imagined becoming a walking repository of verse.

I decided to make memorizing a part of my daily routine. Like flossing. Except I was actually going to do it. Each morning, after waking up and having my coffee, but before reading the newspaper or showering or even putting on proper clothes, I sat down behind my desk and tried to spend ten to fifteen minutes working through a poem.

The problem was that I wasn't any good at it. When I sat down and tried to fill a memory palace with Lewis Carroll's "Jabberwocky," a twenty-eight-line poem composed almost entirely of nonsense words, I couldn't figure out how to transform the "brillig" and "slithy toves" into images, and ended up just memorizing the poem by rote, which was exactly what I wasn't supposed to be doing. Next I tried T. S. Eliot's "The Love Song of J. Alfred Prufrock," a poem I'd always adored, and which I already knew in bits and pieces. "In the room the women come and go / Talking of Michelangelo." How could I forget that? Or rather, how was I supposed to remember it? Was I meant to put an image of women, coming and going, speaking of Michelangelo in my uncle's bathroom? And what was that supposed to look like? Or was I supposed to form an image of women, an image of coming, an image of going, and an image of Michelangelo? I was confused. And this was taking forever. These memory techniques, which had seemed so promising while I'd huddled numb-fingered with Ed on a boulder in Central Park, weren't working out nearly so well now that I was alone in my parents' basement. I felt like I had tried on a slick pair of sneakers at the store, and now that I'd worn them home, I had blisters. Clearly I was missing something.

I turned to my newly acquired copy of the *Rhetorica ad Herennium* and opened to the section that discusses the memorization of words. I was hoping it might offer some hints as to why I was failing so badly, but all the two-thousand-year-old book could provide was consolation. Memorizing poetry and prose is extraordinarily difficult, the author willingly concedes. But that's exactly the point. He explains that learning texts is worth doing not because it's easy but because it's hard. "I believe that they who wish to do easy things without trouble and toil must previously have been trained in more difficult things," he writes.

. . .

Having begun to futz around with memory techniques, I didn't yet have any sense of the true scope of the enterprise I was embarking upon. I still thought of my project as a harmlessly casual experiment. All I wanted to know was whether I really could improve my memory, and, if so, by how much. I certainly hadn't taken Tony Buzan's challenge to try to compete in the USA Memory Championship seriously. After all, there were more than three dozen American mental athletes who trained each year for the event, which takes place every March in New York City. There was no reason to think a journalist who occasionally forgets his own Social Security number could compete against America's top memory geeks. But, as I would soon learn, Americans on the international memory circuit are like Jamaicans on the international bobsledding circuit—easily the most laid-back folks at any competition, and possibly even the most stylish, but on the international stage, we are behind the curve in terms of both technique and training.

Even though the best American mnemonists can memorize hundreds of random digits in an hour, USA records still pale in comparison to those of the Europeans. Generally, nobody in North America takes memory sport seriously enough to stop drinking three months before the world championship, like the eight-time world memory champ Dominic O'Brien used to do, and from the looks of it, few competitors engage in the rigorous physical training regimen that Buzan recommends. (One of his first, unsolicited pieces of advice to me was to get in shape.) Nobody downs daily glasses of cod liver oil or takes omega-3 supplements. Only one American, the four-time national champion Scott Hagwood, has ever been inducted into the KL7.

Even though America has run its national memory championship for as long as any country in the world, the best American memorizer has only finished in the top five of the world championship once, in 1999. Perhaps it says something about our national character that America has produced none of the world's best competitive memorizers—that we're not as detail-obsessed as the Germans, as punctilious as the Brits, or as driven as the Malaysians. Or maybe, as one European soberly suggested to me, Americans have impoverished memories because we are preoccupied with the future, while folks on the other side of the Atlantic are more concerned with the past. Whatever the reason, it became clear that if I wanted to learn more about the art of memory—if I wanted to study with the best in the world—I was going to have to go to Europe.

Having spent several weeks struggling with mixed success to furnish my memory palaces with poetry, I thought it time to enlist some help in order to take my efforts to the next level. The granddaddy of events on the yearlong international memory circuit, the World Memory Championship, was going to be held in Oxford, England, at the end of the summer. I decided I needed to go, and convinced *Discover* magazine to send me to write an article about the competition. I called up Ed to ask if I could crash at his place. Oxford was his home turf—where he'd grown up, gone to college, and now lived at home with his parents on their country estate located on the town's outskirts, in a seventeenth-century stone house called the Mill Farm.

When I arrived at the Mill Farm (or simply "the Milf," as Ed sometimes referred to it) on a sunny summer afternoon a few days before the World Memory Championship, Ed greeted me and carried my bags up to his bedroom, the same one he grew up in, with clothes scattered about the floor and nine decades of cricket almanacs on his bookshelves. Then he took me into the house's oldest wing, a four-hundred-year-old converted stone barn linked to the kitchen. There was a piano in the corner and colorful fabrics draped from the ceiling, the remnants of a party held years ago that were never taken down. At one end of the room was a long wooden table with eight decks of playing cards arranged at the head.

"This is where I practice," Ed said, and pointed to a balcony that jutted into the upper part of the barn. "Images of binary digits come pouring down those stairs over there, right across the room. This is exactly where you'd expect a memory champ to exercise, isn't it?"

Before dinner, an old childhood friend of Ed's named Timmy stopped by to say hello. Ed and I came downstairs to find him at the table chatting with Ed's mother and father, Teen and Rod, while his youngest sister, Phoebe, chopped vegetables from the garden at the kitchen island. Timmy now ran an online application development company. He had driven over in a BMW, wore a crisp polo shirt, and had a warm tan.

Teen introduced me and explained, with a wry laugh, that Ed was my memory coach. Timmy seemed not to believe that Ed was still toying with all this memory stuff. Hadn't it been quite some time since he'd taken that crazy trip to Kuala Lumpur?

"Edward, are you at all nervous that your new student will surpass you?" asked Teen, mostly it seemed for the sake of ribbing her son.

"I don't think anyone needs to be too concerned about that," I said.

"Well, I think it'd strike a tremendous blow for education," said Ed proudly.

"Do you think you could give Ed a nine-to-five job?" Rod asked Timmy.

Ed laughed. "Yes, you know, maybe I could give memory training courses to your employees."

"You could do programming," offered Teen.

"I don't know how to program."

"Your father could teach you."

Rod made a small fortune in the 1990s designing computer software, and retired at an early age to a life of leisure and eccentric pursuits. He is a practicing apiarist and gardener and would like to take the Mill Farm off the electrical grid by exercising his ancient water rights and installing a hydroelectric generator in the creek that runs by the house. Teen teaches developmentally disabled kids at a local school and is an avid reader and tennis player. She is mostly tolerant of Ed's eccentricities, but also cautiously hopeful that Ed might someday direct his considerable talents in a more focused, and perhaps even socially useful, direction.

"What about the law, Edward?" she asked.

"I consider the law to be a zero-sum game, and therefore a pointless use of a life," said Ed. "Being good at being a lawyer means merely, on average, maximizing injustice." Ed leaned over to me. "I used to be quite a promising young man when I was eighteen."

This prompted Phoebe to chime in: "More like thirteen."

While Ed was in the bathroom, I asked Rod if he would be disappointed if his son ended up becoming the next Tony Buzan, a fantastically wealthy self-help guru. Rod pondered the question for a few seconds and stroked his chin. "I think I'd prefer if he became a barrister."

The next morning, at the examination hall at Oxford University, which was hosting the world's finest mnemonists, Ed was sprawled out across a leather sofa, wearing a bright yellow cap and a T-shirt with the words "Ed Kicks Ass—220" emblazoned in bold letters across his chest, above a menacing ironed-on photograph of himself, a cartoon of a karate kick, and a photograph of thonged female hindquarters. (In addition to communicating an intimidating bit of trash talk to his opponents, he explained, those three words, "Ed Kicks Ass," are a mnemonic that helps him remember the number 220.) He was smoking a cigarette (he doesn't take the physical training part of the sport too seriously), and warmly greeting each of the competitors as they strolled through the door. He informed me that since we'd last seen each other, he'd taken an indefinite leave of absence from his PhD program in Paris to pursue "other projects." He also told me that his and Lukas's big plans for the Oxford Mind Academy had been temporarily derailed when, not long after the U.S. championship, Lukas badly seared his lungs in a fire-breathing stunt gone wrong.

Memory championships can be pathologically competitive events, and Ed described his vanity T-shirt as part of a "campaign of pretend intimidation" with the aim of "generally upping the quality of banter between competitors—especially with the Germans." To that end, he had showed up at the championship bearing copies of a cheeky one-page stats sheet that he was handing out to the press and fellow competitors. It described his character (in the third person)—"Irreverent, flamboyant, ready for anything (especially yesterday)"—and his training regime—"Early Rise, Yoga, Skipping, Superfoods (including blueberries and cod liver oil), Four-hours training, two glasses of wine per day (from the potassium rich soil of the Languedoc-Roussillon in Southern France), 30 minutes reflection period at sunset each evening, keeping a journal online." It noted that his "unique abilities" include lucid dreaming and tantric sex. It also described Tony Buzan as "a champion ball-room dancer and a mentor for me throughout puberty," and his thoughts on the future of competitive memory: "Hoping it will be an Olympic sport before 2020," when he is "planning to retire to a life of synaesthesia and senility." His plans for after the championship: "Revolutionizing Western Education."

Sitting on the couch next to him was the legendary world memory champion Ben Pridmore, a man who until that moment I had known only through Google and myth. (I had heard he could memorize playing cards as fast as he could turn them over.) Ben wore a worn-out "One

Fish, Two Fish, Red Fish, Blue Fish" Dr. Seuss T-shirt with a badly stretched collar, and a fanny pack. He was also sporting an enormous wide-brimmed black Australian steer-hide undertaker's hat that he professed to have worn every day for the last six years. "It's my gimmick," he said softly. "It's part of my soul." At his feet there sat a pink and black backpack with the words "Pump It Up" graffitied on the back. He informed us that there were twenty-two decks of playing cards inside, which he intended to memorize the next day in a single hour.

With his bald head, dark beard, face-swallowing glasses, and wide, searching eyes, Ben seemed almost like a figure out of an R. Crumb cartoon. He even had the same shrugged shoulders and loopy strut. The soles of his tattered leather shoes slapped under his feet like flip-flops. He spoke with a gentle, slightly nasal Yorkshire accent, which turned "my" into "me." "I hate me voice," he said when explaining why he'd been so cagey about returning my phone calls during the previous weeks. One of the first pieces of information about himself that he shared with me was that he believed he was England's youngest college dropout. "I was admitted to Kingston on Thames University when I was seventeen, but I dropped out after six months. Now I'm twenty-eight, which is a bit depressing. I'm starting to feel like the old man of memory sports. You know, I was one of the hot newcomers once."

Bad luck does seem to stalk Ben. He'd had no intentions of being at the World Memory Championship. Instead, he had devoted the last six months to memorizing the first 50,000 digits of the mathematical constant pi, which he planned to recite at the Mind Sports Olympiad, a weeklong festival of board games to be held a week after the World Memory Championship. It would have been a new world record. But an obscure Japanese mnemonist named Akira Haraguchi had emerged from nowhere to memorize 83,431 digits just a month earlier.

It took him sixteen hours and twenty-eight minutes to recite them. Ben read about the accomplishment on the Internet and was forced to reevaluate his plans. Instead of trying to learn another 33,432 digits, he gave up and rededicated himself to defending his title as world memory champion. He had spent virtually every free moment of the last six weeks cleaning out memory palaces that had been devoted to pi, undoing months of hard work so that he could reuse the palaces in the memory championships.

Most of the mental athletes on the memory circuit came to the sport the same way I did: They once saw someone perform an outrageous memory stunt, thought it was cool, learned the trick behind it, and then went home and tried it themselves. But Ben missed one critical step. He'd seen someone memorizing playing cards and thought it was cool, and went home and tried it himself. But nobody ever told him how it was done. Without using any techniques at all, he just stared at the cards over and over again until they'd become imprinted on his brain. And the amazing thing is, he kept doing this in his spare time for several months, under the assumption that eventually he'd surely get good at it. He finally got his time down to fifteen minutes using pure rote memorization, a feat in many ways more impressive than his world record time of thirty-two seconds using techniques. It wasn't until he showed up at his first World Memory Championship in 2000 that he found out about the memory palace. After the first day of events wrapped up (he finished near last place), he went to a bookstore, bought one of Tony Buzan's books, decided this was something he had a talent for, and forgot about all of his other extracurricular interests, including his lifelong quest to watch every one of the 1,001 theatrically released Warner Bros. cartoons made between 1930 and 1968.

Ben had been working on a book called "How to Be Clever," which teaches readers how to calculate the day of the week for any date in history, how to memorize a deck of cards, and how to scam an IQ test. "The book is about making people think you're brainy without actually increasing your intelligence," he told me. "The problem is I haven't written very much because I always have more important things to do, like watch cartoons. If I tried to write a serious book on how to improve your life, I'd be rubbish at it, because I haven't got the faintest idea how to improve my life."

The favorite to take Ben's title at the world championship was Dr. Gunther Karsten, the balding, angular forty-three-year-old godfather of German memory sport, who had won every German national contest since 1998. Gunther showed up wearing what I learned is his standard uniform: an imposing pair of black earmuffs and metallic sunglasses whose insides have been completely taped over except for two small pinholes. "Extraneous stimuli," as Gunther calls them, are the memorizer's bête noir. (A retired Danish mnemonist used to compete wearing horse blinders.) He also wore a gold belt buckle embossed with his initials, a gold chain over his tight white T-shirt, and black sailor pants that flared at the bottom. Gunther informed me that in college he was a photo model for Nissan cars, and depending on how you squinted, he looked like the villain in a James Bond movie or an aging figure skater. He was in terrific physical shape, and was, I would soon learn, a fierce competitor. Despite the fact that one of his legs is slightly shorter than the other (from a childhood bone disease), he regularly races in—and wins—track events for middle-aged men. He was carrying around with him a locked, shiny metal suitcase filled with between twenty and thirty decks of playing cards, which he planned to memorize. He wouldn't tell me the exact number for fear it would get back to Ben Pridmore.

The actual competition took place in a large oak-paneled room in one of Oxford's storied old buildings, with tall Gothic windows and oversize portraits of the third Earl of Litchfield and the fourteenth Earl of Derby. The room was arranged no differently than it had been during the school year, when it was used to administer exams to Oxford undergraduates. There were four dozen desks, each of which had a six-inch-tall digital stopwatch clamped to it, which would be used for the last and most exciting event of the contest, speed cards, when the competitors race to commit a single deck of playing cards to memory as fast as possible.

Unlike the U.S. championship, which has just five events, none lasting longer than fifteen minutes, the World Memory Championship is frequently referred to as a "mental decathlon." Its ten events, called "disciplines," span three grueling days, and each tests the competitors' memories in a slightly different way. Contestants have to memorize a previously unpublished poem spanning several pages, pages of random words (record: 280 in fifteen minutes), lists of binary digits (record: 4,140 in thirty minutes), shuffled decks of playing cards, a list of historical dates, and names and faces. Some disciplines, called "speed events," test how much the contestants can memorize in five minutes (record: 405 digits). Two marathon disciplines test how many decks of cards and random digits they can memorize in an hour (records: 2,080 digits and 27 decks of cards).

The first World Memory Championship was held at the posh Athenaeum Club in London in 1991. "I thought, this is insane," recalls Tony Buzan. "We have crossword championships. We have Scrabble championships. We have chess, bridge, poker, draughts, canasta, and Go championships. We have science fair championships. And for the biggest, the most fundamental of all human cognitive processes, memory, there's no championship." He also knew that the idea of a "world memory champion" would be an irresistible draw for the media, and a savvy way to promote his books on mind training.

With the help of his friend Raymond Keene, a British chess grand master who writes the daily chess column for *The Times* (London), Buzan sent out letters to a handful of people who he knew were involved in memory training, and ran an ad in *The Times* advertising the contest. Seven people showed up, including a psychiatric nurse named Creighton Carvello who had memorized the telephone number of every Smith in the Middlesbrough phonebook and another person named Bruce Balmer who had set a record for memorizing two thousand foreign words in a single day. Several of the competitors wore tuxedoes.

The contestants no longer adhere to such a strict dress code, but everything else about the championship has gotten far more serious since 1991. What began as a one-day contest has now expanded to fill an entire weekend. Of all the disciplines in a three-day memory decathlon, the first one of the first day, the poem, is the most universally dreaded. Because of my own faltering efforts to memorize poetry, it was the one event that I wanted to watch most closely. Every year Gunther lobbies to have the event stricken from the contest, or at least replaced with rules that are more—as he puts it—"objective." But poetry is where memorization began, and to cut it from the championship because a few of the competitors find it difficult would run counter to the competition's underlying premise that memorization is a creative and humanizing endeavor. So every year, a new, previously unpublished poem is commissioned for the world championship. For the first few years of the competition, in the early nineties, the poem was written by the British poet laureate Ted Hughes, whom Tony Buzan describes as "an old friend." Since Hughes's death in 1998, the poem has been written by Buzan himself. This year's

108-line free-verse offering, titled "Miserare," came from a collection titled "Requiem for Ted." It began:

With most things in the Universe
I am happy:
Supernovas
The Horse Head Nebula
The Crab
The light-years-big clouds
That are the Womb of Stars

It went on to list the many things Tony Buzan was happy about, including "God's freezing balls," and ended:

I am <u>not</u> happy
That Ted
Is Dead.

The competitors had fifteen minutes to memorize as many lines as possible, and then a half hour to write them on a blank sheet of paper. In order to receive full credit for a line, it had to be rendered perfectly, down to each capital letter and punctuation mark. Competitors who failed to underscore just how "<u>not</u> happy" the author was or who mistakenly thought that Ted was "dead" without a capital D would get only half the total points for that line.

The question of how best to memorize a piece of text, or a speech, has vexed mnemonists for millennia. The earliest memory treatises described two types of recollection: *memoria rerum* and *memoria verborum*, memory for things and memory for words. When approaching a text or a speech, one could try to remember the gist, or one could try to remember verbatim. The Roman rhetoric teacher

Quintilian looked down on *memoria verborum* on the grounds that creating such a vast number of images was not only inefficient, since it would require a gargantuan memory palace, but also unstable. If your memory for a speech hinged on knowing every word, then not only did you have a lot more to remember, but if you forgot a single word, you could end up trapped in a room of your memory palace staring at a blank wall, lost and unable to move on.

Cicero agreed that the best way to memorize a speech is point by point, not word by word, by employing *memoria rerum*. In his *De Oratore*, he suggests that an orator delivering a speech should make one image for each major topic he wants to cover, and place each of those images at a locus. Indeed, the word "topic" comes from the Greek word *topos*, or place. (The phrase "in the first place" is a vestige from the art of memory.)

Perfect recall of words is something our brains simply aren't very good at, a fact famously illustrated in the congressional Watergate hearings of 1973. In his testimony before the Senate Watergate Investigating Committee, President Richard Nixon's counsel John Dean reported to the congressmen on the contents of dozens of meetings related to the cover-up of the break-in. To the president's chagrin and the committee's delight, Dean was able to repeat verbatim many conversations that had taken place in the Oval Office. His recollections were so detailed and seemingly so precise that reporters took to calling him "the human tape recorder." At the time, it hadn't yet been revealed that there had been an actual tape recorder in the Oval Office recording the conversations that Dean had reconstructed from memory.

While the rest of the country took note of the political implications of those tape recordings, the psychologist Ulric Neisser saw them as a valuable data trove. Neisser compared the transcripts of the recordings with Dean's testimony, and analyzed what Dean's memory got right and what it got wrong. Not only did Dean not remember specific quotes correctly—that is to say, *verborum*—he often didn't even properly remember the gist of what had been discussed—*rerum*. But even when his memories were wrong in isolated episodes, notes Neisser, "there is a sense in which he was altogether right." The major themes of his testimony were all accurate: "Nixon wanted the cover-up to succeed; he was pleased when it went well; he was troubled when it began to unravel; he was perfectly willing to consider illegal activities if they would extend his power or confound his enemies." John Dean did not misrepresent, argues Neisser; he did get the details wrong, but he got the important stuff right. We all do the same thing when we try to recount conversations, because without special training our memories tend to only pay attention to the big picture.

It makes sense that our brains would work like that. The brain is a costly organ. Though it accounts for only 2 percent of the body's mass, it uses up a fifth of all the oxygen we breathe, and it's where a quarter of all our glucose gets burned. The brain is the most energetically expensive piece of equipment in our body, and has been ruthlessly honed by natural selection to be efficient at the tasks for which it evolved. One might say that the whole point of our nervous system, from the sensory organs that feed information to the glob of neurons that interprets it, is to develop a sense of what is happening in the present and what will happen in the future, so that we can respond in the best possible way. Strip away the emotions, the philosophizing, the neuroses, and the dreams, and our brains, in the most reductive sense, are fundamentally prediction and planning machines. And to work efficiently, they have to find order in the chaos of possible memories. From the vast amounts of data pouring in through the senses, our brains must quickly sift out which information is likely to have some bearing on the future, attend to that, and ignore the noise. Much of the chaos that our brains filter out is words, because more often than not, the actual language that conveys an idea is just window dressing. What matters is the *res*, the meaning of those words. And that's what our brains are so good at remembering. In real life, it's rare that anyone is asked to recall *ad verbum* outside of congressional depositions and the poetry event at an international memory competition.

Until the last tick of history's clock, cultural transmission meant oral transmission, and poetry, passed from mouth to ear, was the principle medium of moving information across space and from one generation to the next. Oral poetry was not simply a way of telling lovely or important stories, or of flexing the imagination. It was, argues the classicist Eric Havelock, "a massive repository of useful knowledge, a sort of encyclopedia of ethics, politics, history, and technology which the effective citizen was required to learn as the core of his educational equipment." The great oral works transmitted a shared cultural heritage, held in common not on bookshelves, but in brains.

Professional memorizers have existed in oral cultures throughout the world to transmit that heritage through the generations. In India, an entire class of priests was charged with memorizing the Vedas with perfect fidelity. In pre-Islamic Arabia, people known as *Rawis* were often attached to poets as official memorizers. The Buddha's teachings were passed down in an unbroken chain of oral tradition for four centuries until they were committed to writing in Sri Lanka in the first century B.C. And for centuries, a group of human tape recorders called *tannaim* (literally, "reciters") memorized the oral law on behalf of the Jewish community.

The most famous of the Western tradition's oral works, and the first to have been systematically studied, were Homer's *Odyssey* and *Iliad*. These two poems—possibly the first to have been written down in the Greek alphabet—had long been held up as literary archetypes. However, even as they were celebrated as the models to which all literature should aspire, Homer's masterworks had also long been the source of scholarly unease. The earliest modern critics sensed that they were somehow qualitatively different from everything that came after—even a little strange. For one thing, both poems were oddly repetitive in the way they referred to characters. Odysseus was always "clever Odysseus." Dawn was always "rosy-fingered." Why would someone write like that? Sometimes the epithets seemed completely off-key. Why call the murderer of Agamemnon "blameless Aegisthos"? Why refer to "swift-footed Achilles" even when he was sitting down? Or to "laughing Aphrodite" even when she was in tears? In terms of both structure and theme, the *Odyssey* and *Iliad* were also oddly formulaic, to the point of predictability. The same narrative units—gathering armies, heroic shields, challenges between rivals—pop up again and again, only with different characters and different circumstances. In the context of such finely spun, deliberate masterpieces, these quirks seemed hard to explain.

At the heart of the unease about these earliest works of literature were two fundamental questions: First, how could Greek literature have been born ex nihilo with two masterpieces? Surely a few less perfect stories must have come before, and yet these two were among the first on record. And second, who exactly was their author? Or was it authors? There were no historical records of Homer, and no trustworthy biography of the man exists beyond a few self-referential hints embedded in the texts themselves.

Jean-Jacques Rousseau was one of the first modern critics to suggest that Homer might not have been an author in the contemporary sense of a single person who sat down and wrote a story and then published it for others to read. In his 1781 *Essay on the Origin of Languages*, the Swiss philosopher suggested that the *Odyssey* and *Iliad* might have been "written only in men's memories. Somewhat later they were laboriously collected in writing"—though that was about as far as his inquiry into the matter went. Also writing in the eighteenth century, an English diplomat and archaeologist named Robert Wood suggested that Homer was illiterate, and that his works had to have been committed to memory. It was a revolutionary theory, but Wood couldn't back it up with a hypothesis that explained how Homer might have pulled off such an astounding mnemonic feat.

In 1795, the German philologist Friedrich August Wolf argued for the first time that not only were Homer's works not *written down* by Homer, but they also weren't even *by* Homer. They were, rather, a loose collection of songs transmitted by generations of Greek bards, and only redacted in their present written form at some later date.

In 1920, an eighteen-year-old scholar named Milman Parry took up the question of Homeric authorship as his master's thesis at the University of California, Berkeley. He suggested that the reason Homer's epics seemed unlike other literature was because they *were* unlike other literature. Parry had discovered what Wood and Wolf had missed: the evidence that the poems had been transmitted orally was right there in the text itself. All those stylistic quirks, including the formulaic and recurring plot elements and the bizarrely repetitive epithets—"clever Odysseus" and "gray-eyed Athena"—that had always perplexed readers were actually like thumbprints left by a potter: material evidence of how the poems had been crafted. They were mnemonic aids that helped the bard(s) fit the meter and pattern of the line, and remember the essence of the poems. The greatest author of antiquity was actually, Parry argued, just "one of a long tradition of oral poets that . . . composed wholly without the aid of writing."

Parry realized that if you were setting out to create memorable poems, the *Odyssey* and the *Iliad* were exactly the kinds of poems you'd create. It's said that clichés are the worst sin a writer can commit, but to an oral bard, they were essential. The very reason that clichés so easily seep into our speech and writing—their insidious memorability—is exactly why they played such an important role in oral storytelling. And the *Odyssey* and *Iliad*, excuse the cliché, are riddled with them. In a culture dependent on memory, it's critical, in the words of Walter Ong, that people "think memorable thoughts." The brain best remembers things that are repeated, rhythmic, rhyming, structured, and above all easily visualized. The principles that the oral bards discovered, as they sharpened their stories through telling and retelling, were the same basic mnemonic principles that psychologists rediscovered when they began conducting their first scientific experiments on memory around the turn of the twentieth century: Words that rhyme are much more memorable than words that don't; concrete nouns are easier to remember than abstract nouns; dynamic images are more memorable than static images; alliteration aids memory. A striped skunk making a slam dunk is a stickier thought than a patterned mustelid engaging in athletic activity.

The most useful of all the mnemonic tricks employed by the bards was song. As anyone who has ever found himself chanting "By Mennen!" can attest, if you can turn a set of words into a jingle, they can become exceedingly difficult to knock out of your head.

Finding patterns and structure in information is how our brains extract meaning from the world, and putting words to music and rhyme are a way of adding extra levels of pattern and structure to language. It's the reason Homeric bards sang their epic oral poems, the reason that printed Torahs are marked up with little musical notations, and the reason we teach kids the alphabet in a song and not as twenty-six individual letters. Song is the ultimate structuring device for language.

After moving to Harvard and becoming an assistant professor, Parry took an unconventional turn in his work. Rather than hunkering down with old Greek texts, the young classicist took off for Yugoslavia in search of the last bards who still practiced a form of oral poetry resembling the Homeric arts. He returned to Cambridge with thousands of recordings that formed the basis for a new branch of academic research into oral traditions.

In his fieldwork, Parry found that rather than transmitting the text itself from bard to bard and generation to generation, the contemporary Balkan rhapsodists (presumably like their ancient Homeric predecessors) would impart a set of formulaic rules and constraints that allowed the bard—any bard—to reconstruct the poem each time he recited it. Each retelling of the story was not exactly like the one that came before, but it was close.

When the Slavic bards were asked whether they repeated their songs exactly, they responded, "Word for word, and line for line." And yet when recordings of two performances were held up against each other, they clearly were different. Words changed, lines moved around, passages disappeared. The Slavic bards weren't being overconfident; they simply had no concept of verbatim recall. Not that this should have been surprising. Without writing, there is no way to check whether something has been repeated exactly.

The variability that is built into the poetry of oral traditions allows the bard to adapt the material to the audience, but it also allows more memorable versions of the poem to arise. Folklorists have compared oral poems to pebbles worn down by the water. They're made smooth over many retellings as the harder-to-remember pieces get chipped away, or made easier to retain and repeat. Irrelevant digressions are forgotten. Long or rare words are avoided. Between imagery, alliteration, and having to fit the meter of the line, the epic bard usually doesn't have that many possible words to choose from. The structure writes the poem. Indeed, work by Parry's successors has found that virtually every word in the *Odyssey* and the *Iliad* fits into some sort of schema, or pattern, that made the poems easier to remember.

. . .

It's no coincidence that the art of memory was supposedly invented by Simonides at exactly the moment when the use of writing was on the rise in ancient Greece, around the fifth century B.C. Memory was no longer something that could be taken for granted, as it had been during Greece's preliterate epoch. The old techniques of the Homeric bards, of rhythm and formula, were no longer adequate to holding in mind the new and complex thoughts that people were beginning to think. "The original oral performance with its poetry was stripped of functional purpose and relegated to the secondary role of entertainment, one which it always had but which now became its sole purpose," writes Havelock. No longer burdened by the requirements of oral transmission, poetry was free to become art.

By the time the author of the *Ad Herennium* sat down to compose his handbook on oration in the first century B.C., writing was already a centuries-old craft, as fundamental a part of the Roman world as computers are a part of our own. The poems produced by his contemporaries—Virgil, Horace, and Ovid all wrote their masterworks within a century of the *Ad Herennium*—lived on the page. Each word was painstakingly selected, the product of a single artist expressing his singular vision. And once set down, those words were considered inviolable.

The anonymous author of the *Ad Herrenium* suggests that the best method for remembering poetry *ad verbum* is to repeat a line two or three times before trying to see it as a series of images. This is more or less the method that Gunther Karsten uses in the poem competition. He assigns every single word to a route point. But this method has a glaring problem: There are lots of words that simply can't be visualized. What does an "and" look like? Or a "the"? Some two thousand years ago, Metrodorus of Scepsis, a Greek contemporary of Cicero's, offered a solution to the quandary of how to see the unseeable. Metrodorus developed a system of shorthand images that would stand in for conjunctions, articles, and other syntactical connectors. It allowed him to memorize anything he read or heard verbatim. Indeed, Metrodorus's library of symbols seems to have been widely used in ancient Greece. The *Ad Herennium* mentions that "most of the Greeks who have written on memory have taken the course of listing images that correspond to a great many words, so that persons who wished to learn these images by heart would have them ready without expending effort in search of them." Though Gunther doesn't use Metrodorus's symbols, which unfortunately have been lost to history, he has created his own dictionary of images for each of the two hundred most common words that can't easily be visualized. "And" is a circle ("and" rhymes with *rund*, which means round in German). "The" is someone walking on his knees (*die*, a German word for "the," rhymes with *Knie*, the German word for "knee"). When the poem reaches a period, he hammers a nail into that locus.

Gunther could just as easily be memorizing a VCR repair manual as a Shakespearean sonnet. In fact, a VCR repair manual would probably be a good deal easier, since it is filled with concrete, easily visualized words like "button," "television," and "plug." The challenge of memorizing poetry is its abstractness. What do you do with words like "ephemeral" or "self" that are impossible to *see*?

Gunther's method of creating an image for the un-imageable is a very old one: to visualize a similarly sounding, or punning, word in its place. The fourteenth-century English theologian and mathematician Thomas Bradwardine, who was later appointed archbishop of Canterbury, took this kind of verbatim memorization to its highest and most absurd level of development. He described a means of *memoria sillabarum*, or "memory by syllables," which could be used to memorize words that were hard to visualize. Bradwardine's system involved breaking the word into its constituent syllables and then creating an image for each syllable based on another word that begins with that syllable. For example, if one wanted to remember the syllable "ab-," one would picture an abbot. For "ba-" one might visualize a crossbowman (a *balistarius*). When strung together, a chain of these syllables becomes a kind of rebus puzzle. (The Swedish pop group Abba could be recalled as an abbot getting shot by a crossbow.) This process of transforming words into images involves a kind of remembering by forgetting: In order to memorize a word by its sound, its meaning has to be completely dismissed. Bradwardine could translate even the most pious benediction into a preposterous scene. To remember the topic sentence of a sermon that begins "Benedictus Dominus qui per," he'd see "the sainted Benedictine dancing to his left with a white cow with super-red teats who holds a partridge, while with his right hand he either mangles or caresses St. Dominic."

The art of memory was, from its origins, always a bit risqué. Preoccupied with Gothic and sometimes downright lewd imagery, it was bound to come in for harsh criticism from the prudes eventually. It's amazing, in a way, that the casual marriage of the reverent and irreverent that Bradwardine practiced in his imagination was not more upsetting to some of the more priggish clergy. When the moralistic attack finally came, it was led by the sixteenth-century Puritan reverend William Perkins of Cambridge. He decried the art of memory as idolatrous and "impious, because it calls up absurd thoughts, insolent, prodigious, and the like which stimulate and light up depraved carnal affections." Carnal indeed. Perkins was particularly steamed by Peter of Ravenna's admission that he used the lustful image of a young woman to excite his memory.

Of the ten events in the World Memory Championship, the poem has bred the greatest number of different strategies. But broadly speaking, mental athletes take two general tacks, which happen to segregate the pool of competitors fairly neatly by gender. While Gunther and most of the other men on the circuit subscribe to a methodical strategy, the women tend to approach the challenge in a more emotional way. Fifteen-year-old Corinna Draschl, an Austrian in a red T-shirt and matching red socks and red baseball cap, told me she can't memorize a text unless she understands what it means. Even more than that, she has to understand how it feels. She breaks the poem into small chunks and then assigns a series of emotions to each short segment. Rather than associate the words with images, she associates them with feelings.

"I feel how the writer feels, what he is meaning. I imagine whether he's happy or sad," she told me in the hallway outside the competition hall. This is not dissimilar from how actors are taught to memorize scripts. Many actors will tell you that they break their lines into units they call "beats," each of which involves some specific intention or goal on the character's part, which they train themselves to empathize with. This technique, known as Method acting, was pioneered in Russia by Konstantin Stanislavski around the turn of the last century. Stanislavski was interested in these techniques not for their mnemonic potential but rather as tools to help the actor more realistically depict his character. But Method acting is a technique for giving a line more associational hooks to hang on by embedding it in a context of both emotional and physical cues. Method acting is a way of making words memorable. Indeed, studies have found that if you ask someone to memorize a sentence like "Pick up a pen," it's much more likely to stick if the person literally picks up a pen as they're learning the sentence.

Ultimately, Gunther ended up losing the poetry event to Corinna Draschl, and losing the championship as well. The top prize went to one of his protégés, a quiet and intensely focused eighteen-year-old Bavarian law student named Clemens Mayer, who spoke only choppy

English and made it clear that he had no interest in practicing the language on me. After botching the spoken numbers and names-and-faces events, Ben Pridmore landed in fourth place overall, lowered the brim of his black hat, and walked out the door alone, vowing that he would begin preparing the next day to reclaim his title one year hence.

Ed fared even worse. Of the three dozen competitors, he was one of only eleven who failed to memorize an entire deck of cards in either of the two speed cards trials, which is like a place kicker missing an extra point twice in a row. He'd been gunning for an especially low time that would take him to the upper ranks, but he'd lost control and burned too hard. He ended up finishing a disappointing eleventh place overall, and sulked out the door, sodden with sweat. I ran after him and grabbed him to ask what had happened. "Too much ambition," was all he would say, shaking his head. "I'll see you back at the house."

He walked across the Magdalen Bridge to go find a pub where he could watch some cricket and drink Guinness until he'd forgotten his failure.

Standing at the front of the Oxford examination hall, watching the competitors scratch their heads and twiddle their pens as they struggled to recall "Miserare," I felt acutely aware of how odd it was that we've come to this: that the only place left where the ancient art of memory is being practiced, or at least celebrated, is in this rarefied competition, and among this quirky subculture. Here in one of the world's most storied centers of learning were the last vestiges of a glorious Golden Age of Memory.

It is hard not to feel as though a tremendous devolution has taken place between that Golden Age and our own comparatively leaden one. People used to labor to furnish their minds. They invested in the acquisition of memories the same way we invest in the acquisition of things. But today, beyond the Oxford examination hall's oaken doors, the vast majority of us don't trust our memories. We find shortcuts to avoid relying on them. We complain about them endlessly, and see even their smallest lapses as evidence that they're starting to fail us entirely. How did memory, once so essential, end up so marginalized? Why did these techniques disappear? How, I wondered, did our culture end up forgetting how to remember?

SEVEN

THE END OF REMEMBERING

Once upon a time, there was nothing to do with thoughts except remember them. There was no alphabet to transcribe them in, no paper to set them down upon. Anything that had to be preserved had to be preserved in memory. Any story that would be retold, any idea that would be transmitted, any piece of information that would be conveyed, first had to be remembered.

Today it often seems we remember very little. When I wake up, the first thing I do is check my day planner, which remembers my schedule so that I don't have to. When I climb into my car, I enter my destination into a GPS device, whose spatial memory supplants my own. When I sit down to work, I hit the play button on a digital voice recorder or open up a notebook that holds the contents of my interviews. I have photographs to store the images I want to remember, books to store knowledge, and now, thanks to Google, I rarely have to remember anything more than the right set of search terms to access humankind's collective memory. Growing up, in the days when you still had to punch seven buttons, or turn a clunky rotary dial, to make a telephone call, I could recall the numbers of all my close friends and family. Today, I'm not sure if I know more than four phone numbers by heart. And that's probably more than most. According to a survey conducted in 2007 by a neuropsychologist at Trinity College Dublin, fully a third of Brits under the age of thirty can't remember even their *own* home land line number without pulling it up on their handsets. The same survey found that 30 percent of adults can't remember the birthdays of more than three immediate family members. Our gadgets have eliminated the need to remember such things anymore.

Forgotten phone numbers and birthdays represent minor erosions of our everyday memory, but they are part of a much larger story of how we've supplanted our own natural memory with a vast superstructure of technological crutches—from the alphabet to the BlackBerry. These technologies of storing information outside our minds have helped make our modern world possible, but they've also changed how we think and how we use our brains.

In Plato's *Phaedrus*, Socrates describes how the Egyptian god Theuth, inventor of writing, came to Thamus, the king of Egypt, and offered to bestow his wonderful invention upon the Egyptian people. "Here is a branch of learning that will . . . improve their memories," Theuth said to the Egyptian king. "My discovery provides a recipe for both memory and wisdom." But Thamus was reluctant to accept the gift. "If men learn this, it will implant forgetfulness in their souls," he told the god. "They will cease to exercise their memory and become forgetful; they will rely on that which is written, calling things to remembrance no longer from within themselves, but by means of external marks. What you have discovered is a recipe not for memory, but for reminding. And it is no true wisdom that you offer your disciples, but only its semblance, for by telling them of many things without teaching them anything, you will make them seem to know much, while for the most part they will know nothing. And as men filled not with wisdom but with the conceit of wisdom, they will be a burden to their fellow-men."

Socrates goes on to disparage the idea of passing on his own knowledge through writing, saying it would be "singularly simple-minded to believe that written words can do anything more than remind one of what one already knows." Writing, for Socrates, could never be anything more than a cue for memory—a way of calling to mind information already in one's head. Socrates feared that writing would lead the culture down a treacherous path toward intellectual and moral decay, because even while the quantity of knowledge available to people might increase, they themselves would come to resemble empty vessels. I wonder if Socrates would have appreciated the flagrant irony: It's only because his pupils Plato and Xenophon put his disdain for the written word into written words that we have any knowledge of it today.

Socrates lived in the fifth century B.C., at a time when writing was ascendant in Greece, and his own views were already becoming antiquated. Why was he so put off by the idea of putting pen to paper? Securing memories on the page would seem to be an immensely superior way of retaining knowledge compared to trying to hold it in the brain. The brain is always making mistakes, forgetting, misremembering. Writing is how we overcome those essential biological constraints. It allows our memories to be pulled out of the fallible wetware of the brain and secured on the less fallible page, where they can be made permanent and (one sometimes hopes) disseminated far, wide, and across time. Writing allows ideas to be passed across generations, without fear of the kind of natural mutation that is necessarily a part of oral traditions.

To understand why memory was so important in the world of Socrates, we have to understand something about the evolution of writing, and how different early books were in both form and function. We have to go back to an age before printing, before indexes and tables of contents, before the codex parceled texts into pages and bound them at the edge, before punctuation marks, before lowercase letters, even before there were spaces between words.

Today we write things down precisely so we don't have to hold them in our memories. But through at least the late Middle Ages, books were thought of not merely as replacements for memory, but also as memory aids. As Thomas Aquinas put it, "Things are written down in material books to help the memory." One read in order to remember, and books were the best available tools for impressing information into the mind. In fact, manuscripts were sometimes copied for no reason other than to help their copier memorize them.

In the time of Socrates, Greek texts were written on long, continuous scrolls—some stretching up to sixty feet—pasted together from sheets of pressed papyrus reeds imported from the Nile Delta. These texts were cumbersome to read, and even more cumbersome to write. It would be tough to invent a less user-friendly way of accessing information. In fact, it wasn't until about 200 B.C. that the most basic punctuation marks were invented by Aristophanes of Byzantium, the director of the Library of Alexandria, and all they consisted of was a single dot at either the bottom, middle, or top of the line letting readers know how long to pause between sentences. Instead, words ran together in an unending stream of capital letters known as *scriptio continua*, broken up by neither spaces nor punctuation. Words that started on one line would spill over onto the next without even a hyphen.

ASYOUCANSEEITSNOTVERYEASYTOREADTE
XTWRITTENWITHOUTSPACESORPUNCTUATI
ONOFANYKINDOREVENHELPFULLYPOSITIO

NEDLINEBREAKSANDYETTHISWASEXACTLY
THEFORMOFINSCRIPTIONUSEDINANCIENT
GREECE

Unlike the letters in this book, which form words that carry semantic meaning, letters written in *scriptio continua* functioned more like musical notes. They signified the sounds that were meant to come out of one's mouth. Reconstituting those sounds into discrete packets of words that could be understood first required hearing them. And just as it is difficult for all but the most gifted musicians to read musical notes without actually singing them, so too was it difficult to read texts written in *scriptio continua* without speaking them aloud. In fact, we know that well into the Middle Ages, reading was an activity generally carried out aloud, a kind of performance, and one most often given before an audience. "Lend ears" is a phrase often repeated in medieval texts. When St. Augustine, in the fourth century A.D., observed his teacher St. Ambrose reading to himself without moving his tongue or murmuring, he thought the unusual behavior so noteworthy as to record it in his *Confessions*. It was probably not until about the ninth century, around the same time that spacing became common and the catalog of punctuation marks grew richer, that the page provided enough information for silent reading to become common.

The difficulties associated with reading such texts meant that there was a very different relationship between reading and memory than the one we know today. Since sight-reading *scriptio continua* was difficult, reciting a text aloud with fluency required a reader to have a degree of familiarity with it. He—and it was mostly he's—had to prepare with it, punctuate it in his mind, memorize it—in part, if not in full—because turning a string of sounds into meaning was not something you could do easily on the fly. The text had to be learned before it could be performed. After all, the way one punctuated a text written in *scriptio continua* could make all the difference in the world. As the historian Jocelyn Penny Small pointed out, GODISNOWHERE comes out a lot differently when rendered as GOD IS NOW HERE versus GOD IS NOWHERE.

What's more, a scroll written in *scriptio continua* had to be read top to bottom if anything was to be taken from it. A scroll has just a single point of entry, the first word. Because it has to be unwound to be read, and because there are no punctuation marks or paragraphs to break up the text—to say nothing of page numbers, a table of contents, chapter divisions, and an index—it is impossible to find a specific piece of information without scanning the whole thing, head to toe. It is not a text that can be easily consulted—until it is memorized. This is a key point. Ancient texts couldn't be readily scanned. You couldn't pull a scroll off the shelf and quickly find a specific excerpt unless you had some baseline familiarity with the entire text. The scroll existed not to hold its contents externally, but rather to help its reader navigate its contents internally.

One of the last places where this tradition of recitation still survives is in the reading of the Torah, an ancient handwritten scroll that can take upward of a year to inscribe. The Torah is written without vowels or punctuation (though it does have spaces, an innovation that came to Hebrew before Greek), which means it's extremely difficult to sight-read. Though Jews are specifically commanded not to recite the Torah from memory, there's no way to read a section of the Torah *without* having invested a lot of time familiarizing yourself with it, as any once-bar-mitzvahed boy can tell you. I can personally vouch for this. On the day I became a man, I was really just a parrot in a yarmulke.

Though years of language use condition us not to notice, *scriptio continua* has more in common with the way we actually speak than the artificial word divisions on this page. Spoken sentences flow together seamlessly as one long, blurry drawn-out sound. We don't speak with spaces. Where one word ends and another begins is a relatively arbitrary linguistic convention. If you look at a sonographic chart visualizing the sound waves of someone speaking English, it's practically impossible to tell where the spaces are, which is one of the reasons why it's proven so difficult to train computers to recognize speech. Without sophisticated artificial intelligence capable of figuring out context, a computer has no way of telling the difference between "The stuffy nose may dim liquor" and "The stuff he knows made him lick her."

For a period, Latin scribes actually did try separating words with dots, but in the second century A.D., there was a reversion—a giant and very curious step backward, it would seem—to the old continuous script used by the Greeks. Spaces weren't seen again in Western writing for another nine hundred years. From our vantage point today, separating words seems like a no-brainer. But the fact that it was tried and rejected says a lot about how people used to read. So, too, does the fact that the ancient Greek word most commonly used to signify "to read" was *ánagignósko*, which means to "know again," or "to recollect." Reading as an act of remembering: From our modern vantage point, could there be a more unfamiliar relationship between reader and text?

Today, when we live amid a deluge of printed words—would you believe that more than a million new books were printed last year?—it's hard to imagine what it must have been like to read in the age before Gutenberg, when a book was a rare and costly handwritten object that could take a scribe months of labor to produce. Even as late as the fourteenth century, there might be just several dozen copies of any given text in existence, and those copies might well be chained to a desk or lectern in some library, which, if it contained a hundred other books, would have been considered particularly well stocked. If you were a medieval scholar reading a book, you knew that there was a reasonable likelihood you'd never see that particular text again, and so a high premium was placed on remembering what you read. You couldn't just pull a book off the shelf to consult it for a quote or an idea. For one thing, modern bookshelves with their rows of outward-facing spines hadn't even been invented yet. That didn't happen until sometime around the sixteenth century. For another thing, books still tended to be heavy, hardly portable objects. It was only in the thirteenth century that bookmaking technology advanced to the point that the Bible could be compiled in a single volume rather than a collection of independent books, and yet it still weighed more than ten pounds. And even if you did happen to have a text you needed close at hand, the chances of finding what you were looking for without reading the whole thing start to finish were slim. Indexes were not yet common, nor were page numbers or tables of contents.

But these gaps were gradually filled. And as the book itself changed, so too did the crucial role of memory in reading. By about the year 400, the parchment codex, with its leaves of pages bound at the spine like a modern hardcover, had all but completely replaced scrolls as the preferred way to read. No longer did a reader have to unfurl a long document to find a passage. A reader could just turn to the appropriate page.

The first concordance of the Bible, a grand index that consumed the labors of five hundred Parisian monks, was compiled in the thirteenth century, around the same time that chapter divisions were introduced. For the first time, a reader could refer to the Bible without having previously memorized it. One could find a passage without knowing it by heart or reading the text all the way through. Soon after the concordance, other books with alphabetical indexes, page numbers, and tables of contents began to appear, and as they did, they again helped change the essence of what a book was.

The problem of the book before the index and table of contents is that for all the material contained in a scroll or between the covers of a book, it was impossible to navigate. What makes the brain such an incredible tool is not just the sheer volume of information it contains but the ease and efficiency with which it can find that information. It uses the greatest random-access indexing system ever invented—one that computer scientists haven't come even close to replicating. Whereas an index in the back of a book provides a single address—a page number—for each important subject, each subject in the brain has hundreds if not thousands of addresses. Our internal memories are associational, nonlinear. You don't need to know where a particular memory is stored in order to find it. It simply turns up—or doesn't—when you need it. Because of the dense network that interconnects our memories, we can skip around from memory to memory and idea to idea very rapidly. From Barry White to the color white to milk to the Milky Way is a long voyage conceptually, but a short jaunt neurologically.

Indexes were a major advance because they allowed books to be accessed in the nonlinear way we access our internal memories. They helped turn the book into something like a modern CD, where you can skip directly to the track you want, as compared to unindexed books, which, like cassette tapes, force you to troll laboriously through large swaths of material in order to find the bit you're looking for. Along with page numbers and tables of contents, the index changed what a book was, and what it could do for scholars. The historian Ivan Illich has argued that this represented an invention of such magnitude that "it seems reasonable to speak of the pre- and post-index Middle Ages." As books became easier and easier to consult, the imperative to hold their contents in memory became less and less relevant, and the very notion of what it meant to be erudite began to evolve from possessing information internally to knowing where to find information in the labyrinthine world of external memory.

. . .

To our memory-bound predecessors, the goal of training one's memory was not to become a "living book," but rather a "living concordance," a walking index of everything one had read, and all the information one had acquired. It was about more than merely possessing an internal library of facts, quotes, and ideas; it was about building an organizational scheme for accessing them. Consider, for example, Peter of Ravenna, a leading fifteenth-century Italian jurist (also, one gets the impression, one of the fifteenth century's leading self-promoters) who authored one of the era's most successful books on memory training. Titled *Phoenix*, it was translated into several languages and reprinted all across Europe. It was just the most famous of a handful of memory treatises created from the thirteenth century onward that helped make memory techniques that had long been the exclusive purview of scholars and monks available to a wider audience of doctors, lawyers, tradesmen, and everyday folks who just wanted to remember stuff. One finds books from the period on every variety of mnemonic subject, including how to use the art of memory in gambling, how to use it to keep track of debts, how to memorize the contents of ships, how to remember the names of acquaintances, and how to memorize playing cards. Peter, for his part, bragged of having memorized twenty thousand legal points, a thousand texts by Ovid, two hundred of Cicero's speeches and sayings, three hundred sayings of philosophers, seven thousand texts from Scripture, as well as a host of other classical works.

For leisure, he would reread books cached away in his many memory palaces. "When I left my country to visit as a pilgrim the cities of Italy, I can truly say I carried everything I owned with me," he wrote. To store all those images, Peter started with a hundred thousand loci, but he was always picking up new memory palaces on his travels across

Europe. He constructed a mental library of sources and quotations on every important subject, classified alphabetically. He boasts, for example, that filed away in his brain under the letter A were sources on the subjects "*de alimentis, de alienatione, de absentia, de arbitris, de appellationibus, et de similibus quae jure nostro habentur incipientibus in dicta littera A*"—"about provisions, about foreign property, about absence, about judges, about appeals, and about similar matters in our law which begin with the letter A." Each piece of knowledge was assigned a specific address. When he wished to expound on a given topic, he simply reached into the proper chamber of the proper memory palace and pulled out the proper source.

When the point of reading is, as it was for Peter of Ravenna, remembering, you approach a text very differently than most of us do today. Now we put a premium on reading quickly and widely, and that breeds a kind of superficiality in our reading, and in what we seek to get out of books. You can't read a page a minute, the rate at which you're probably reading this book, and expect to remember what you've read for any considerable length of time. If something is going to be made memorable, it has to be dwelled upon, repeated.

In his essay "The First Steps Toward a History of Reading," Robert Darnton describes a switch from "intensive" to "extensive" reading that occurred as books began to proliferate. Until relatively recently, people read "intensively," says Darnton. "They had only a few books—the Bible, an almanac, a devotional work or two—and they read them over and over again, usually aloud and in groups, so that a narrow range of traditional literature became deeply impressed on their consciousness."

But after the printing press appeared around 1440, things began gradually to change. In the first century after Gutenberg, the number of books in existence increased fourteenfold. It became possible, for the first time, for people without great wealth to have a small library in their own homes, and a trove of easily consulted external memories close at hand.

Today, we read books "extensively," without much in the way of sustained focus, and, with rare exceptions, we read each book only once. We value quantity of reading over quality of reading. We have no choice, if we want to keep up with the broader culture. Even in the most highly specialized fields, it can be a Sisyphean task to try to stay on top of the ever-growing mountain of words loosed upon the world each day.

Few of us make any serious effort to remember what we read. When I read a book, what do I hope will stay with me a year later? If it's a work of nonfiction, the thesis, maybe, if the book has one. A few savory details, perhaps. If it's fiction, the broadest outline of the plot, something about the main characters (at least their names), and an overall critical judgment about the book. Even these are likely to fade. Looking up at my shelves, at the books that have drained so many of my waking hours, is always a dispiriting experience. *One Hundred Years of Solitude*: I remember magical realism and that I enjoyed it. But that's about it. I don't even recall when I read it. About *Wuthering Heights* I remember exactly two things: that I read it in a high school English class and that there was a character named Heathcliff. I couldn't say whether I liked the book or not.

I don't think I'm an exceptionally bad reader. I suspect that many people, maybe even most, are like me. We read and read and read, and we forget and forget and forget. So why do we bother? Michel de Montaigne expressed the dilemma of extensive reading in the sixteenth century: "I leaf through books, I do not study them," he wrote. "What I retain of them is something I no longer recognize as anyone else's. It is only the material from which my judgment has profited, and the thoughts and ideas with which it has become imbued; the author, the place, the words, and other circumstances, I immediately forget."

He goes on to explain how "to compensate a little for the treachery and weakness of my memory," he adopted the habit of writing in the back of every book a short critical judgment, so as to have at least some general idea of what the tome was about and what he thought of it.

You might think that the advent of printing, and the ability to more easily offload memories from brains onto paper, would have immediately rendered the old memory techniques irrelevant. But that's not what happened. At least not right away. In fact, paradoxically, at exactly the moment when a neat rendering of history would suggest that the art of memory should have been on its way to obsolescence, it underwent its greatest renaissance.

Ever since Simonides, the art of memory had been about creating architectural spaces in the imagination. But in the sixteenth century, an Italian philosopher and alchemist named Giulio Camillo—known as "Divine Camillo" to his many admirers and "the Quack" to his many detractors—had the clever idea of making concrete what had for the previous two thousand years always been an ethereal idea. It occurred to him that the system would work a whole lot better if someone transformed the metaphor of the memory palace into a real wooden building. He imagined creating a "Theater of Memory" that would serve as a universal library containing all the knowledge of mankind. It may sound like the premise of a Borges story, but it was a very real project, with very real backers, and it made Camillo into one of the most famous men in all of Europe. King Francis I of France made Camillo promise that the secrets of his theater would never be revealed to anyone but him, and invested five hundred ducats toward its completion.

Camillo's wooden memory palace was shaped like a Roman amphitheater, but instead of the spectator sitting in the seats looking down on the stage, he stood in the center and looked up at a round, seven-tiered edifice. All around the theater were paintings of Kabbalistic and mythological figures as well as endless rows of drawers and boxes filled with cards, on which were printed everything that was known, and—it was claimed—everything that was knowable, including quotations from all the great authors, categorized according to subject. All you had to do was meditate on an emblematic image and the entirety of knowledge stored in that section of the theater would be called immediately to mind, allowing you to "be able to discourse on any subject no less fluently than Cicero." Camillo promised that "by means of the doctrine of loci and images, we can hold in the mind and master all human concepts and all the things that are in the entire world."

That was a grand claim, and with hindsight, sure, it sounds like hocus-pocus. But Camillo was convinced that there existed a set of magical symbols that could organically represent the entire cosmos. Just as the image of the she-male represented the concept of e-mailing in that first memory palace I built to house Ed's to-do list, Camillo believed there were images that could encapsulate vast and powerful concepts about the universe, and simply by memorizing those images, one would be able understand the hidden connections underlying everything.

A scale wooden model of Camillo's theater was exhibited in Venice and Paris, and hundreds—perhaps thousands—of cards were drafted to fill the theater's boxes and drawers. The artists Titian and Salviati were enlisted to paint the theater's symbolic imagery. However, that seems to be about as far as things got. The theater was never actually completed, and all that remains of the grand scheme is a short, posthumously published manifesto, *The Idea of the Theater*, dictated on his deathbed over the course of a week. Written in the future tense without any images or diagrams, it is, to put it mildly, a confusing book.

Though history had largely forgotten the man who promised the ultimate technology for remembering—"divine" lost out to "quack" in almost every assessment—Camillo's reputation was resurrected in the twentieth century thanks to the efforts of the historian Frances Yates, who helped reconstruct the theater's blueprints in her book *The Art of Memory*, and the Italian literature professor Lina Bolzoni, who has helped explain how Camillo's theater was more than just the work of a nut job, but actually the apotheosis of an entire era's ideas about memory.

The Renaissance, with its fresh translations of ancient Greek texts, brought about a renewed fascination with Plato's old idea that there is a transcendental ideal reality of which our own world is but a pale shadow. In Camillo's Neoplatonic vision of the universe, images in the mind were a way of accessing that ideal realm, and the art of memory was a secret key to unlocking the occult structure of the universe. Memory was transformed from a tool of rhetoric, as it had been for the ancients, or an instrument of pious meditation, as it had been for the medieval scholastic philosophers, into a purely mystical art.

Even more than Camillo, the greatest practitioner of this dark, mystical form of mnemonics was the Dominican friar Giordano Bruno. In his book *On the Shadow of Ideas*, published in 1582, Bruno promised that his art "will help not only the memory but also all the powers of the soul." Memory training, for Bruno, was the key to spiritual enlightenment.

Bruno had literally come up with a new twist on the old art of memory. Drawing inspiration from the palindromically named thirteenth-century Catalan philosopher and mystic Ramon Llull, Bruno invented a device that would allow him to turn any word into a unique image. Bruno imagined a series of concentric wheels, each of which had 150 two-letter pairs around its perimeter, corresponding to all of the combinations that could be formed by the thirty letters of the alphabet (the twenty-three letters of classical Latin, plus seven Greek and Hebrew letters that didn't have an equivalent in the Latin alphabet) and the five vowels: AA, AE, AI, AO, AU, BA, BE, BO, etc. On the innermost wheel, the 150 two-letter combinations were each paired with a different mythological or occult figure. On the perimeter of the second wheel were 150 actions and predicaments—"sailing," "on the carpet," "broken"—corresponding to another set of letter pairs. The third wheel consisted of 150 adjectives, the fourth wheel had 150 objects, and the fifth wheel had 150 "circumstances," such as "dressed in pearls" or "riding a sea monster." By properly aligning the wheels, any word up to five syllables long could be translated into a unique, vivid image. For example, the word *crocitus*, Latin for "croaking of a raven," becomes an image of the Roman diety "Pilumnus advancing rapidly on the back of a donkey with a bandage on his arm and a parrot on his head." Bruno was convinced that his opaque and divinely loopy invention was a major step forward for the arts of memory, analogous in scale, he proclaimed, to the technological leap from carving letters in trees to the printing press.

Bruno's scheme, tinged with magic and the occult, deeply troubled the church. His unorthodox ideas, which included such heresies as a belief in Copernican heliocentrism and a conviction that Mary wasn't really a virgin, ultimately landed him in the unforgiving arms of the Inquisition. In 1600, he was burned at the stake in the Campo dei Fiori in Rome and his ashes dispersed in the Tiber River. Today, a statue of Bruno stands in the plaza where he was immolated, a beacon to free-thinkers and mental athletes the world over.

Once the Enlightenment had finally put to bed the Renaissance's obsession with occult memory theaters and Llullian wheels, the art of memory passed into a new but no less harebrained era—the age of the "get smart quick" scheme—which to this day it hasn't yet escaped. Over a hundred treatises on mnemonics were published in the nineteenth century, with titles like "American Mnemotechny" and "How to Remember." They bear a conspicuous resemblance to the memory improvement books that can be found in the self-help aisle at bookstores today. The most popular of these nineteenth-century mnemonic handbooks was written by Professor Alphonse Loisette, an American "memory doctor" who, despite his prolific remembering, "had somehow forgotten that he was born Marcus Dwight Larrowe and that he had no degree," as one article notes. The fact that I was able to find 136 used copies of Loisette's 1886 book *Physiological Memory: The Instantaneous Art of Never Forgetting* selling for as little as $1.25 on the Internet is evidence of its once immense popularity.

Loisette's book is essentially a collection of mnemonic systems for remembering sundry trivia, like the order of American presidents, the counties of Ireland, the Morse telegraphic alphabet, the British territorial regiments, and the names and uses of the cranial nerves. Loisette claimed his system was wholly unrelated to classical mnemonics, for which he professed disdain, and that he had discovered, entirely by himself, the "laws of natural memory."

Loisette charged as much as twenty-five dollars (more than five hundred dollars in today's money) to impart this knowledge to his pupils in seminars held all across the country, including classes at just about every prestigious university on the eastern seaboard. Inductees into the "Loisette System" were made to sign a contract binding them to secrecy, with a penalty of five hundred dollars (over ten thousand dollars in today's money) should they divulge the professor's methods. There was, it seems, good money to be made peddling secrets of memory improvement to a credulous American audience. According to the doctor's own numbers, he earned today's equivalent of almost a half million dollars over a single fourteen-week stretch in the winter of 1887.

In 1887, Samuel L. Clemens, better known as Mark Twain, first crossed paths with Loisette and enrolled in a memory course lasting several weeks. Twain used to say that his "memory was never loaded with anything but blank cartridges," and had long had an interest in memory improvement. He came out of the course a deep believer in Loisette's system. In fact, he was so taken with Loisette that he independently published a broadside claiming that ten thousand dollars an hour would be a bargain for the invaluable tricks the doctor was imparting. He would later regret this testimonial, but not until after it found its way onto virtually every piece of printed matter Loisette produced.

In 1888, G. S. Fellows, out of "that keen sense of justice and innate love of liberty, characteristic of every true American" published a book called *"Loisette" Exposed* that set out to clarify that "Professor" "Loisette"—yes, both appellations bore their own set of scare quotes—was both a "humbug and a fraud." The 224-page book revealed that his methods were either ripped off and repackaged from older sources or else obscenely oversold. Surely Loisette's humbuggery and fraudulence ought to have been self-evident to someone as versed in the ways of the world as Mark Twain, but Twain was a profligate fad chaser, and always interested in the next big thing. (His personal investment of $300,000—$7 million today—in the Paige typesetter, an early competitor of the Linotype, was only the most ruinous of several ambitious projects he poured his money into.)

Twain himself was continually experimenting with new memory techniques to aid him on the lecture circuit. At one point early in his career, he wrote the first letter of topics he planned to drop into his speech on each of his ten fingernails, but that never really worked, since audiences began to suspect him of having some sort of vain interest in his hands. During the summer of 1883, while he was writing *Huckleberry Finn*, Twain procrastinated by developing a game to teach his children the English monarchs. It worked by mapping out the lengths of their reigns using pegs along a road near his home. Twain was essentially turning his backyard into a memory palace. In 1885, he patented "Mark Twain's Memory Builder: A Game for Acquiring and Retaining All Sorts of Facts and Dates." Twain's notebooks are filled with pages dedicated to his spatial memory game.

Twain imagined national clubs organized around his mnemonic game, regular newspaper columns, a book, and international competitions with prizes. He became convinced that the entire corpus of historical and scientific facts that any American student needed to know could be taught through his ingenious invention. "Poets, statesmen, artists, heroes, battles, plagues, cataclysms, revolutions . . . the invention of the logarithm, the microscope, the steam-engine, the telegraph—anything and everything all over the world—we dumped it all in among the English pegs," he wrote in his 1899 essay "How to Make History Dates Stick." Unfortunately, like the Paige typesetter, the game turned out to be a financial bust, and Twain was eventually forced to abandon it. He wrote to his friend the novelist William Dean Howells, "If you haven't ever tried to invent an indoor historical game, don't."

Like so many before him, Twain had gotten swept up in the promise of vanquishing forgetfulness. He had drunk of the same wacky elixir that had intoxicated Camillo and Bruno and Peter of Ravenna, and his story should probably be read as a cautionary tale to anyone embarking on a course of memory training. Perhaps, in retrospect, the resemblances between Dr. Loisette and today's memory gurus should have sent me running for the hills. And yet they didn't.

Twain lived in an age when the technologies for storing and retrieving external memories—paper, books, the recently invented photograph and phonograph—were still primitive compared to what we have today. He could not have foreseen how the proliferation of digital information at the beginning of the twenty-first century would hasten the pace at which our culture has become capable of externalizing its memories. With our blogs and tweets, digital cameras, and unlimited-gigabyte e-mail archives, participation in the online culture now means creating a trail of always present, ever searchable, unforgetting external memories that only grows as one ages. As more and more of our lives move online, more and more is being captured and preserved in ways that are dramatically changing the relationship between our internal and external memories. We are moving toward a future, it seems, in which we will have all-encompassing external memories that record huge swaths of our daily activity.

I was convinced of this by a seventy-three-year-old computer scientist at Microsoft named Gordon Bell. Bell sees himself as the vanguard of a new movement that takes the externalization of memory to its logical extreme: a final escape from the biology of remembering.

"Each day that passes I forget more and remember less," writes Bell in his book *Total Recall: How the E-Memory Revolution Will Change Everything*. "What if you could overcome this fate? What if you never had to forget anything, but had complete control over what you remembered—and when?"

For the last decade, Bell has kept a digital "surrogate memory" to supplement the one in his brain. It ensures that a record is kept of anything and everything that might be forgotten. A miniature digital camera, called a SenseCam, dangles around his neck and records every sight that passes before his eyes. A digital recorder captures every sound he hears. Every phone call placed through his landline gets taped and every piece of paper Bell reads is immediately scanned into his computer. Bell, who is completely bald, often smiling, and wears rectangular glasses and a black turtleneck, calls this process of obsessive archiving "lifelogging."

All this obsessive recording may seem strange, but thanks to the plummeting price of digital storage, the increasing ubiquity of digital sensors, and better artificial intelligence to sort through the mess of data we're constantly collecting, it's becoming easier and easier to capture and remember ever more information from the world around us. We may never walk around with cameras dangling from our necks, but Bell's vision of a future in which computers remember everything that happens to us is not nearly as absurd as it might at first sound.

Bell made his name and fortune as an early computing pioneer at the Digital Equipment Corporation in the 1960s and '70s. (He's been called the "Frank Lloyd Wright of computers.") He's an engineer by nature, which means he sees problems and tries to build solutions. With the SenseCam, he is trying to fix an elemental human problem: that we forget our lives almost as fast as we live them. But why should any memory fade when there are technological solutions that can preserve it?

In 1998, with the help of his assistant Vicki Rozyki, Bell began backfilling his lifelog by systematically scanning every document in the dozens of banker boxes he'd amassed since the 1950s. All of his old photos, engineering notebooks, and papers were digitized. Even the logos on his T-shirts couldn't escape the scanner bed. Bell, who had always been a meticulous preservationist, figures he's probably scanned and thrown away three quarters of all the stuff he's ever owned. Today his lifelog takes up 170 gigabytes, and is growing at the rate of about a gigabyte each month. It includes over 100,000 e-mails, 65,000 photos, 100,000 documents, and 2,000 phone calls. It fits comfortably on a hundred-dollar hard drive.

Bell can pull off some sensational stunts with his "surrogate memory." With his custom search engine, he can, in an instant, figure out where he was and whom he was with at any moment in time, and then, in theory, check to see what that person said. And because he's got a photographic record of everywhere he's ever been and everything he's ever seen, he has no excuse for ever losing anything. His digital memory never forgets.

Photographs, videos, and digital recordings are, like books, prosthetics for our memories—chapters in the long journey that began when the Egyptian god Theuth came to King Thamus and offered him the gift of writing as a "recipe for both memory and wisdom." Lifelogging is the logical next step. Maybe even the logical final step, a kind of *reductio ad absurdum* of a cultural transformation that has been slowly unfolding for millennia.

I wanted to meet Bell and see his external memory at work. His project would seem to offer the ultimate counterargument to all the effort I was investing in training my internal memory. If we're bound to have computers that never forget, why bother having brains that remember?

When I visited his immaculate Microsoft Research office overlooking the San Francisco Bay, Bell wanted to show me how he uses his external memory to help find things that have gone missing in his internal memory. Because memories are associative, finding the odd misplaced fact is often an act of triangulation. "The other day, I was trying to find a house I had looked at online," Bell told me, leaning back in his chair. "All I could remember about it is that I was talking to the realtor on the phone at the time." He pulled up a time line of his life on his computer, found the phone conversation on it, and then immediately pulled up all the Web sites he was looking at while he was on the phone. "I call them information barbs," says Bell. "All you need is to remember a hook." The more barbs there are stored in one's digital memory, the easier it is to find what you're looking for.

Bell has a wealth of external memories at his fingertips. By far the biggest problem Bell faces is how to avoid the fate of Funes and S and keep from drowning in a sea of meaningless trivia. So much of remembering happens at the moment of encoding, because we only tend to remember what we pay attention to. But Bell's lifelog pays attention to everything. "Don't ever filter, and never throw anything away" is his motto.

"Do you ever feel burdened by the sheer volume of memories you're collecting?" I asked him.

He scoffed at the notion. "No way. I feel this is tremendously freeing."

The SenseCam is not a beautiful machine. It's a black box, about the size of a pack of cigarettes, that dangles around Bell's neck. Inconspicuous it's not. But then again, the first computers took up entire rooms and the earliest cell phones were the size of cinder blocks. It doesn't take much imagination to see how future versions of the SenseCam could be embedded in a pair of eyeglasses, or inconspicuously sewn into clothing, or even somehow tucked under the surface of the skin or embedded in a retina.

For now, Bell's internal and external memories don't mesh seamlessly. In order for him to access one of his stored external memories, he still has to find it on his computer and "re-input" it into his brain through his eyes and ears. His lifelog may be an extension of him, but it's not yet a part of him. But is it so far-fetched to believe that at some point in the not-too-distant future the chasm between what Bell's computer knows and what his mind knows may disappear entirely? Eventually, our brains may be connected directly and seamlessly to our lifelogs, so that our external memories will function and feel as if they are entirely internal. And of course, they will also be connected to the greatest external memory repository of all, the Internet. A surrogate memory that recalls everything and can be accessed as naturally as the memories stored in our neurons: It would be the decisive weapon in the war against forgetting.

This may sound like science fiction, but already cochlear implants can convert sound waves directly into electrical impulses and channel them into the brain stem, allowing previously deaf people to hear. In fact, they've already been installed in more than 200,000 human heads. And primitive cognitive implants that create a direct interface between the brain and computers have already allowed paraplegics and patients with ALS (Lou Gehrig's disease) to control a computer cursor, a prosthetic limb, even a digital voice simply through the force of thought. These neuroprosthetics, which are still highly experimental and have been implanted in only a handful of patients, essentially wiretap the brain, and allow direct communication between man and machine. The next step is a brain-computer interface that lets the mind exchange data directly with a digital memory bank, a project that a few cutting-edge researchers are already working on, and which is bound to become a major area of research in the decades ahead.

You don't have to be a reactionary, a fundamentalist, or a Luddite to wonder whether plugging brains into computers and seamlessly merging internal and external memory would ultimately be such a terrific idea. Today bioethicists work up sweats over such hot-potato topics as genetic engineering and neurotropic "cognitive steroids," but these kinds of enhancements are just tweaking the dials compared with what it would mean to fully marry our internal and external memories. A smarter, taller, stronger, disease-resistant person who lives to 150 is still, in the end, just a person. But if we could give someone a perfect memory and a mind that taps directly into the entire collective knowledge of humanity, well, that's when we might need to consider expanding our vocabulary.

But perhaps instead of thinking of these memories as externalized or off-loaded—as categorically different from memories that reside in the brain—we should view them as *extensions* of our internal memories. After all, even internal memories are accessible only by degrees.

There are events and facts I know I know, but I don't know how to find. Even if I can't immediately recall where I celebrated my seventh birthday or the name of my second cousin's wife, those facts may nevertheless be lurking somewhere in my brain, waiting for the right cue to pop back into consciousness, in just the same way that all the facts in Wikipedia are lurking just a mouse click away.

We Westerners tend to think of the "self," the elusive essence of who we are, as if it were some starkly delimited entity. Even if modern cognitive neuroscience has rejected the old Cartesian idea of a homuncular soul that resides in the pineal gland and controls the human body, most of us still believe there is a distinct "me" somewhere up there driving the bus. In fact, what we think of as "me" is almost certainly something far more diffuse and hazier than it's comfortable to contemplate. At the least, most people assume that their self could not possibly extend beyond the boundaries of their epidermis into books, computers, a lifelog. But why should that be the case? Our memories, the essence of our selfhood, are actually bound up in a whole lot more than the neurons in our brain. At least as far back as Socrates's diatribe against writing, our memories have always extended beyond our brains and into other storage containers. Bell's lifelogging project simply brings that truth into focus.

EIGHT

THE OK PLATEAU

If you visited my office in the fall of 2005, you would have seen a Post-it note—one of my external memories—stuck to the wall above my computer monitor. Whenever my eyes strayed from the screen, I saw the words "Don't Forget to Remember," a gentle reminder that for the next several months until the USA Memory Championship, I needed to strive to replace my regular procrastination patterns with more productive mnemonic exercises. Instead of browsing the Web or walking around the block to cool my eyes, I'd pick up a list of random words and try to memorize it. Rather than take a magazine or book along with me on the subway, I'd whip out a page of random numbers. Did I understand, at the time, how weird I was becoming?

I started trying to use my memory in everyday life, even when I wasn't practicing for the handful of arcane events that would be featured in the championship. Strolls around the neighborhood became an excuse to memorize license plates. I began to pay a creepy amount of attention to name tags. I memorized my shopping lists. I kept a calendar on paper, and also one in my mind. Whenever someone gave me a phone number, I installed it in a special memory palace.

Remembering numbers proved to be one of the real world applications of the memory palace that I relied on almost every day. I used a technique known as the "Major System," invented in the seventeenth century by Johann Winkelmann, which is nothing more than a simple code to convert numbers into phonetic sounds. Those sounds can then be turned into words, which can in turn become images for a memory palace. The code works like this:

0	1	2	3	4
S	T or D	N	M	R
5	6	7	8	9
L	Sh or Ch	K or G	F or V	P or B

The number 32, for example, would translate into MN, 33 would be MM, and 34 would be MR. To make those consonants meaningful, you're allowed to freely intersperse vowels. So the number 32 might turn into an image of a man, 33 could be your mom, and 34 might be the Russian space station Mir. Similarly, the number 86 might be a fish, 40 a rose, and 92 a pen. You might visualize 3,219 as a man (32) playing a tuba (19), or maybe a person from Manitoba (3,219). Likewise, 7,879 would translate to KFKP, which might turn into a single image of a coffee cup, or two images of a calf and a cub. The advantage of the Major System is that it's straightforward, and you can begin using it right out of the box. (When I first learned it, I immediately memorized my credit card and bank account numbers.) But nobody wins any international memory competitions with the Major System.

When it comes to memorizing long strings of numbers, like a hundred thousand digits of pi or the career batting averages of every New York Yankee Hall of Famer, most mental athletes use a more complex technique that is known on the Worldwide Brain Club (the online forum for memory junkies, Rubik's cubers, and mathletes) as "person-action-object," or, simply, PAO. It traces its lineage directly back to the loopy combinatorial mnemonics of Giordano Bruno and Ramon Llull.

In the PAO system, every two-digit number from 00 to 99 is represented by a single image of a person performing an action on an object. The number 34 might be Frank Sinatra (a person) crooning (an action) into a microphone (an object). Likewise, 13 might be David Beckham kicking a soccer ball. The number 79 could be Superman flying with a cape. Any six-digit number, like say 34-13-79, can then be turned into a single image by combining the person from the first number with the action from the second and the object from the third—in this case, it would be Frank Sinatra kicking a cape. If the number were instead 79-34-13, the mental athlete might imagine the equally bizarre image of Superman crooning at a soccer ball. There's nothing inherently Sinatraish about the number 34 or Beckhamesque about 13. Unlike the Major System, those associations are entirely arbitrary, and have to be learned in advance, which is to say it takes a lot of remembering just to be able to remember. There's a big fixed cost in terms of time and effort to compete on the memory circuit. But what makes this system so potent is that it effectively generates a unique image for every number from 0 to 999,999. And because the algorithm necessarily generates unlikely scenes, PAO images tend, by their nature, to be memorable.

The sport of competitive memory is driven by an arms race of sorts. Each year someone—usually it's a competitor who is temporarily underemployed or a student with an unstructured summer vacation—comes up with an ever more elaborate technique for remembering more stuff more quickly, forcing the rest of the field to play catch-up.

Ed had just spent the previous six months developing what he described as "the most elaborate mnemonic behemoth ever brought to bear at a memory championship." His new system, which he referred to as "Millennium PAO," represented an upgrade from the two-digit system used by most European competitors to a three-digit system consisting of a thousand different person-action-object images. It would allow him to convert every number from zero to 999,999,999 into a unique image that would hopefully be impossible to confuse with any other. "While before I had a little two-digit laser boat that could dart through numbers like a tuna on amphetamines, now I have a three-digit sixty-four-gun Man of War," he boasted. "It is enormously powerful, yet potentially difficult to control." If the system worked, he figured, it would advance the sport of competitive memory by a quantum leap.

Mental athletes memorize decks of playing cards in much the same way, using a PAO system in which each of the fifty-two cards is associated with its own person/action/object image. This allows any triplet of cards to be combined into a single image, and for a full deck to be condensed into just eighteen unique images (52 divided by 3 is 17, with one card left over).

With Ed's help, I laboriously created my own PAO system, which involved dreaming up fifty-two separate person/action/object images. To be maximally memorable, one's images have to appeal to one's own sense of what is colorful and interesting. Which means that a mental athlete's stock of PAO images is a pretty good guide to the gremlins that live in someone's subconscious: in my case, 1980s and early 1990s TV icons; in Ben Pridmore's case, cartoon characters; in Ed's case, lingerie models and Depression-era English cricketers. The king of hearts, for me, was Michael Jackson moonwalking with a white glove. The king of clubs was John Goodman eating a hamburger, and the king of diamonds was Bill Clinton smoking a cigar. If I were to memorize the king of hearts, king of clubs, and king of diamonds in order, I would create an image of Michael Jackson eating a cigar. Before I could memorize any decks of cards, I first had to memorize those fifty-two images. No minor job.

But my PAO system pales in comparison to the system that Ben Pridmore uses for cards. In the fall of 2002, he quit the job he'd held for six and a half years as an assistant accountant at a meat factory in Lincolnshire, spent a week in Vegas counting cards, and then came back to England and spent the next six months watching cartoons, getting qualified to teach English as a second language, and developing an entirely new mnemonic nuclear arsenal. Instead of creating a single person-action-object image for each card in the deck, Ben spent dozens of hours dreaming up a unique image for every two-card combination. When he sees the queen of hearts followed by the ace of diamonds, that's a unique image. When he sees the ace of diamonds followed by the queen of hearts, that's a different unique image. That's 52 times 52, or 2,704, possible two-card combinations for which Ben has an image pre-memorized. And like Ed, he places three images at each of his loci. That means he's able to condense an entire pack of cards into just nine loci (52 divided by 6), and twenty-seven packs of cards—the most he's ever been able to memorize in a single hour—into just 234 places.

It's hard to say which is the more admirable component of this feat: Ben's mental or manual dexterity. He has developed an ability to quickly thumb two cards at a time off the top of the deck, in the process spreading them just enough to reveal the suit and number in the corner of both. When he's going at top speed, he looks at each pair of cards for less than a second.

Ben developed a similarly Byzantine system for memorizing binary digits, which enables him to convert any ten-digit-long string of ones and zeros into a unique image. That's 2^{10}, or 1,024, images set aside for binaries. When he sees 1101001001, he immediately sees it as a single chunk, an image of a card game. When he sees 0111011010, he instantaneously conjures up an image of a cinema. In international memory competitions, mental athletes are given sheets of 1,200 binary digits, thirty to a row, forty rows to a page. Ben turns each row of thirty digits into a single image. The number 110110100000111011010001011010, for example, is a muscleman putting a fish in a tin. At the time, Ben held the world record for having learned 3,705 random ones and zeroes in half an hour.

Every mental athlete has a weakness, an Achilles heel. Ben's is names and faces. His scores in the event are always near the bottom of the pack. "I don't tend to look at people's faces when I talk to them," he told me. "In fact, I have no idea what lots of people I know really look like." To get around this problem, he has been developing a new mnemonic system for the event that would assign numerical codes to eye color, skin color, hair color, hair length, nose, and mouth shape. He figures that if people's faces could only be turned into a string of digits, they'd be a cinch to remember.

When I first set out to train my memory, the prospect of learning these elaborate techniques seemed preposterously daunting. But Anders Ericsson and I struck a deal. I would give him the meticulous records of all my training, which would be useful data for his research on expertise. In return Tres and Katy, his graduate students, would analyze that data in search of ways I could perform better. After the memory championship, I had promised to return to Tallahassee for a couple days of follow-up testing so they could get a journal article out of the whole enterprise.

Ericsson has studied the process of skill acquisition from dozens of different angles in almost as many different fields, and if there were any general secrets to becoming an expert, he was the person most likely to reveal them. What I already knew from talking with him extensively, and from reading almost every book and paper he'd written, was that in domain after domain, he'd found a common set of techniques that the most accomplished individuals tend to employ in the process of becoming an expert—general principles of expertise acquisition. Those principles would be my secret weapon.

Over the next several months, while I toiled away with PAO in my parents' basement, Ericsson kept close tabs on my development. I kept him apprised of my evolving thoughts about the impending competition, which I noticed had gradually begun to shift from innocent curiosity to zealous competitiveness. When I'd get stuck, I'd call Ericsson up for advice, and he'd inevitably send me scurrying for some journal article that he promised would help me understand my shortcomings. At one point, a few months into my training, my memory stopped improving. No matter how much I practiced, I couldn't memorize a deck of playing cards any faster. I was stuck in a rut, and I couldn't figure out why. "My card times have hit a plateau," I lamented to him.

"I would recommend you check out the literature on speed typing," he replied.

When people first learn to use a keyboard, they improve very quickly from sloppy single-finger pecking to careful two-handed typing, until eventually the fingers move so effortlessly across the keys that the whole process becomes unconscious and the fingers seem to take on a mind of their own. At this point, most people's typing skills stop progressing. They reach a plateau. If you think about it, it's a strange phenomenon. After all, we've always been told that practice makes perfect, and many people sit behind a keyboard for at least several hours a day in essence practicing their typing. Why don't they just keep getting better and better?

In the 1960s, the psychologists Paul Fitts and Michael Posner attempted to answer this question by describing the three stages that anyone goes through when acquiring a new skill. During the first phase, known as the "cognitive stage," you're intellectualizing the task and discovering new strategies to accomplish it more proficiently. During the second "associative stage," you're concentrating less, making fewer major errors, and generally becoming more efficient. Finally you reach what Fitts called the "autonomous stage," when you figure that you've gotten as good as you need to get at the task and you're basically running on autopilot. During that autonomous stage, you lose conscious control over what you're doing. Most of the time that's a good thing. Your mind has one less thing to worry about. In fact, the autonomous stage seems to be one of those handy features that evolution worked out for our benefit. The less you have to focus on the repetitive tasks of everyday life, the more you can concentrate on the stuff that really matters, the stuff that you haven't seen before. And so, once we're just good enough at typing, we move it to the back of our mind's filing cabinet and stop paying it any attention. You can actually see this shift take place in fMRI scans of people learning new skills. As a task becomes automated, the parts of the brain involved in conscious reasoning become less active and other parts of the brain take over. You could call it the "OK plateau," the point at which you decide you're OK with how good you are at something, turn on autopilot, and stop improving.

We all reach OK plateaus in most things we do. We learn how to drive when we're in our teens and then once we're good enough to avoid tickets and major accidents, we get only incrementally better. My father has been playing golf for forty years, and he's still—though it will hurt him to read this—a duffer. In four decades his handicap hasn't fallen even a point. How come? He reached an OK plateau.

Psychologists used to think that OK plateaus marked the upper bounds of innate ability. In his 1869 book *Hereditary Genius*, Sir Francis Galton argued that a person could only improve at physical and mental activities up until he reached a certain wall, which "he cannot by any education or exertion overpass." According to this view, the best we can do is simply the best we can do.

But Ericsson and his fellow expert performance psychologists have found over and over again that with the right kind of concerted effort, that's rarely the case. They believe that Galton's wall often has much less to do with our innate limits than simply with what we consider an acceptable level of performance.

What separates experts from the rest of us is that they tend to engage in a very directed, highly focused routine, which Ericsson has labeled "deliberate practice." Having studied the best of the best in many different fields, he has found that top achievers tend to follow the same general pattern of development. They develop strategies for consciously keeping out of the autonomous stage while they practice by doing three things: focusing on their technique, staying goal-oriented, and getting constant and immediate feedback on their performance. In other words, they force themselves to stay in the "cognitive phase."

Amateur musicians, for example, are more likely to spend their practice time playing music, whereas pros are more likely to work through tedious exercises or focus on specific, difficult parts of pieces. The best ice skaters spend more of their practice time trying jumps that they land less often, while lesser skaters work more on jumps they've already mastered. Deliberate practice, by its nature, must be hard.

When you want to get good at something, how you spend your time practicing is far more important than the amount of time you spend. In fact, in every domain of expertise that's been rigorously examined, from chess to violin to basketball, studies have found that the number of years one has been doing something correlates only weakly with level of performance. My dad may consider putting into a tin cup in his basement a good form of practice, but unless he's consciously challenging himself and monitoring his performance—reviewing, responding, rethinking, rejiggering—it's never going to make him appreciably better. Regular practice simply isn't enough. To improve, we must watch ourselves fail, and learn from our mistakes.

The best way to get out of the autonomous stage and off the OK plateau, Ericsson has found, is to actually practice failing. One way to do that is to put yourself in the mind of someone far more competent at the task you're trying to master, and try to figure out how that person works through problems. Benjamin Franklin was apparently an early practitioner of this technique. In his autobiography, he describes how he used to read essays by the great thinkers and try to reconstruct the author's arguments according to Franklin's own logic. He'd then open up the essay and compare his reconstruction to the original words to see how his own chain of thinking stacked up against the master's. The best chess players follow a similar strategy. They will often spend several hours a day replaying the games of grand masters one move at a time, trying to understand the expert's thinking at each step. Indeed, the single best predictor of an individual's chess skill is not the amount of chess he's played against opponents, but rather the amount of time he's spent sitting alone working through old games.

The secret to improving at a skill is to retain some degree of conscious control over it while practicing—to force oneself to stay out of autopilot. With typing, it's relatively easy to get past the OK plateau. Psychologists have discovered that the most efficient method is to force yourself to type 10 to 20 percent faster than your comfort pace and to allow yourself to make mistakes. Only by watching yourself mistype at that faster speed can you figure out the obstacles that are slowing you down and overcome them. By bringing typing out of the autonomous stage and back under conscious control, it is possible to conquer the OK plateau.

Ericsson suggested I try the same thing with cards. He told me to find a metronome and to try to memorize a card every time it clicked. Once I figured out my limits, he instructed me to set the metronome 10 to 20 percent faster than that and keep trying at the quicker pace until I stopped making mistakes. Whenever I came across a card that was particularly troublesome, I was supposed to make a note of it, and see if I could figure out why it was giving me problems. It worked, and within a couple days I was off the OK plateau and my card times began falling again at a steady clip.

If they're not practicing deliberately, even experts can see their skills backslide. Ericsson shared with me an incredible example of this. Even though you might be inclined to trust the advice of a silver-haired doctor over one fresh out of medical school, it's been found that in a few fields of medicine, doctors' skills don't improve the longer they've been practicing. The diagnostic accuracy of professional mammographers, for example, doesn't get more accurate over the years. Why would that be?

For most mammographers, practicing medicine is not deliberate practice, according to Ericsson. It's more like putting into a tin cup than working with a coach. That's because mammographers usually only find out if they missed a tumor months or years later, if at all, at which point they've probably forgotten the details of the case and can no longer learn from their successes and mistakes.

One field of medicine in which this is definitively not the case is surgery. Unlike mammographers, surgeons tend to get better with time. What makes surgeons different from mammographers, according to Ericsson, is that the outcome of most surgeries is usually immediately apparent—the patient either gets better or doesn't—which means that surgeons are constantly receiving feedback on their performance. They're always learning what works and what doesn't, always getting better. This finding leads to a practical application of expertise theory: Ericsson suggests that mammographers regularly be asked to evaluate old cases for which the outcome is already known. That way they can get immediate feedback on their performance.

Through this kind of immediate feedback, experts discover new ways to perform ever better and push our collective OK plateaus ever higher. People have been swimming for as long as people have been getting neck-deep in water. You'd think that as a species, we'd have maxed out how fast we could swim long ago. And yet new swimming records are set every year. Humans keep getting faster and faster. "Olympic swimmers from early this century would not even qualify for swim teams at competitive high schools," notes Ericsson. Likewise, "the gold medal performance at the original Olympic marathon is regularly attained by amateurs just to qualify as a participant in the Boston Marathon." And the same is true not just of athletic pursuits, but in virtually every field. The thirteenth-century philosopher Roger Bacon claimed that "nobody can obtain to proficiency in the science of mathematics by the method hitherto known unless he devotes to its study thirty or forty years." Today, the entire body of mathematics known to Bacon is now acquired by your average high school junior.

There's no reason to think that the most talented athletes alive today possess that much more innate talent than the most talented athletes of the past. And there's also no reason to believe that improvements in running shoes or swimwear—while certainly of some significance—could be responsible for the totality of these dramatic improvements. What's changed is the amount and quality of training that athletes must endure to achieve world-class status. The same is true not just of running and swimming, but of javelin throwing, ice skating, and every other athletic pursuit. There is not a single sport in which records don't regularly fall. If there are plateaus out there, collectively we have not reached them yet.

How is it that we continue to surpass ourselves? Part of Ericsson's answer is that the barriers we collectively set are as much psychological as innate. Once a benchmark is deemed breakable, it usually doesn't take long before someone breaks it. For a long time, people thought that no one would ever run a mile in under four minutes. It was considered an immovable barrier, like the speed of light. When Roger Bannister, a twenty-year-old British medical student, finally broke the four-minute mile in 1954, his accomplishment was splashed across the front pages of newspapers around the world and hailed as one of the greatest athletic achievements of all time. But the barrier turned out to be more like a floodgate. It took only six weeks before an Australian named John Landy ran the mile a second and a half faster than Bannister, and within a few years four-minute miles were commonplace. Today, all professional middle-distance runners are expected to clock four-minute miles and the world record has fallen to 3 minutes and 43.13 seconds. At the World Memory Championship, at least half the existing world records fall each year.

Instead of thinking of enhancing my memory as analogous to stretching my height or improving my vision or tweaking some other fundamental attribute of my body, Ericsson encouraged me to think of it more like improving a skill—more like learning to play an instrument.

We usually think about our memory as a single, monolithic thing. It's not. Memory is more like a collection of independent modules and systems, each relying on its own networks of neurons. Some people have good memories for numbers but are always forgetting words; some people remember names but not to-do lists. SF, Ericsson's work-study undergraduate who expanded his digit span tenfold, had not increased some generalized memory capacity. Rather, he'd simply become an expert at digit memorization. When he tried to memorize lists of random consonants, he could still only remember about seven of them.

This, more than anything, is what differentiates the top memorizers from the second tier: They approach memorization like a science. They develop hypotheses about their limitations; they conduct experiments and track data. "It's like you're developing a piece of technology, or working on a scientific theory," the two-time world champ Andi Bell once told me. "You have to analyze what you're doing."

If I would have any chance at catapulting myself to the top tier of the competitive memorization circuit, my practice would have to be focused and deliberate. That meant I needed to collect data and analyze it for feedback. And that meant this whole operation was about to get ratcheted up.

I set up a spreadsheet on my laptop to keep track of how long I was practicing and of any difficulties I was having along the way. I made graphs of everything, and tracked the steady upticks in my scores in a journal:

August 19: Did 28 cards in 2:57.

August 20: Did 28 cards in 2:39. Solid time.

August 24: Did 38 cards in 4:40. Not so good.

September 8: Sitting in a Starbucks procrastinating instead of working on an overdue article. Memorized 46 digits in five minutes . . . Pathetic. Then did 48 cards in 3:32. Decided finally to change my images for the fours. Goodbye female actresses, hello mental athletes. Clubs = Ed Cooke, diamonds = Gunther Karsten, hearts = Ben Pridmore, spades = me.

October 2: Did 70 random words in fifteen minutes. Not good! Lost points because I confused the words "grow" with "growth" and "bicycle" with "bike." From now on, when a word has multiple close variations, make a careful mental note in palace next to the confusing image!

October 16: Just did 87 random words. I'm doing too much eyeing of the clock and glancing around the room instead of memorizing. I'm losing time. Concentrate man, concentrate!

Attention, of course, is a prerequisite to remembering. Generally when we forget the name of a new acquaintance, it's because we're too busy thinking about what we're going to say next, instead of paying attention. Part of the reason techniques like visual imagery and the memory palace work so well is that they enforce a degree of attention and mindfulness that is normally lacking. You can't create an image of a word, a number, or a person's name without dwelling on it. And you can't dwell on something without making it more memorable. The problem I was running into in my training was that I was simply getting bored by it, and allowing my mind's eye to wander. No matter how crude, colorful, and explicit the images one paints in one's memory palaces, one can only look at pages of random numbers for so long before beginning to wonder if there isn't something more interesting going on in another room. Like the sound of putting.

Ed, who had taken to referring to me as "son," "young man," and "Herr Foer," insisted that the cure for my distractibility lay in an equipment upgrade. All serious mnemonists wear earmuffs. A few of the most serious competitors wear blinders to constrict their field of view and shut out peripheral distractions. "I find them ridiculous, but in your case, they may be a sound investment," said Ed on one of our regular twice-weekly phone check-ins. That afternoon, I went out to the hardware store and bought a pair of industrial-grade earmuffs and a pair of plastic laboratory safety goggles. I spray-painted them black and then drilled a small eyehole through each lens. Henceforth I would always wear them to practice.

It was easy enough to explain to people that I was living with my parents to save a few bucks while I cut my teeth as a writer. But what I was doing in their basement, with pages of random numbers taped to the walls and old high school yearbooks (purchased at flea markets) cracked open on the floor, was, if not downright shameful, at least something to lie about.

When my father would visit me in the basement to ask if I'd like to putt with him for a few minutes, I'd quickly hide the page of numbers

I was memorizing and pretend to be diligently at work on something else, like an article that some publication might compensate me for with a check that might in turn be handed over to a landlord. Sometimes I'd take off my earmuffs and memory goggles and turn around to discover that my father had been standing in the doorway, just watching me.

If Ericsson was my professor, Ed had taken on the role of yogi and manager. He set a schedule for me for the next four months, with benchmarks I was supposed to meet along the way, and a strict regimen of half an hour of practice each morning, plus two five-minute booster sessions in the afternoon. A computer program tested me and kept detailed records of my mistakes, so that we could analyze them later. I e-mailed my times to Ed every few days, and he would write back with suggestions about how I could improve.

Eventually, I decided I needed to go back to the Mill Farm to get some more face time with my coach. I scheduled my return trip to England to coincide with Ed's twenty-fifth birthday party, an epic affair that he had been talking up since I had first visited England for the World Memory Championship.

Ed's party was held in the Milf's old stone barn, which Ed had spent the better part of a week converting into an experimental vessel for his philosophy of parties. "I'm trying to find a framework for manipulating conversation, space, movement, mood, and expectation so that I can see how they influence one another," he told me. "In order to track all these parameters, I treat people not as volitional entities but as automata—particles really—which bounce around the party. And as host of the party, I take seriously my responsibility of bouncing them around in the best possible manner."

Glittery textiles hung from the rafters to the floor, dividing the barn into a collection of small rooms. The only way in or out was through a network of tunnels, which could be navigated only by slithering on one's belly. The space under the grand piano was turned into a fort, and a circle was formed around the fireplace out of a collection of raggedy couches that had been stacked on top of tables.

"The people who actually get through the tunnel networks have been through an adventure. They have had to struggle a tiny bit, and therefore upon arrival, they feel a sense of gratitude, relief, and accomplishment, and are committed to the project of having a good experience, with the most possible vigor and imagination. I think your memory training is extremely similar to this. Although it sounds silly to say 'No pain, no gain,' it's true. One has to hurt, to go through a period of stress, a period of self-doubt, a period of confusion. And then out of that mess can flow the richest tapestries."

I crawled behind him through a ten-foot-long pitch-black tunnel and emerged into a room filled neck-deep with balloons. Each room, he explained, was supposed to function like a chamber of a memory palace. His party was designed to be maximally memorable.

"Too often one is just left in a haze about what happened at a party because it's a single, undifferentiated space," he said. "One of the advantages of this kind of setup is that the experiences in each room get kept in each room, and are isolated from other experiences. One leaves the party with a beautiful repertoire of events, upon which one can dwell during old and middle age."

In order to facilitate social interaction, Ed felt it was critical that partygoers not be able to recognize one another. Ben Pridmore, who had taken a four-hour train ride down from Derby, wore a black cape and a terrifying mask of a mohawked man-eater he called Grunch. Lukas Amsüss (recovered from his fire-breathing fiasco), who flew in from Vienna just for the party, arrived wearing a nineteenth-century Austrian military uniform with a sash and medals. One of Ed's old friends from Oxford wore a full-body tiger suit. Another showed up in blackface and dreadlocks. Ed wore a curly wig, a dress, pantyhose, and a generously apportioned bra. In recognition of my being the only Yank at the party, I had my face painted like Captain America.

The highlight of the evening was the card-off. Shortly before midnight, Ed gathered his fifty or so guests in the basement of the barn and announced that in honor of his quarter-century of existence, two of the greatest card memorizers of all time were going to go head-to-head in competition. Ben, still wearing his black cape but no longer his Grunch mask, perched on a beanbag at one end of a long table littered with empty plastic sangria cups and the skeletal remains of an entire lamb that had been spit-roasted over a backyard bonfire. Lukas sat down at the other end of the table in his Austrian military uniform.

"First, I'd like to give the assembled here a few details about these two individuals' capacities to remember packs of cards," Ed announced. "Lukas was one of the first people in the world to break the forty-second barrier for a pack of cards. For a long time in the memory community, which comprises eleven people, this was regarded as the four-minute mile. He busted that mark and busted it again, and was once upon a time the world champion in speed cards. He is also one of the founding members of a distinguished society of memorizers known as the KL7. Of course, his terrific memory would be much better if he weren't perennially drunk," Ed said hyperbolically. Lukas lifted his plastic cup and nodded it in Ed's direction. "You see, Lukas introduced me to an amusing and useful machine that he built with his engineering friends in Vienna, which allows you to drink four glasses of beer in less than three seconds. It's got a valve mechanism they had to purchase off an aerospace company. Unfortunately, Lukas has used it a bit too much recently. He hasn't memorized a deck of cards in almost a year. However, the last time he did it, he recorded a time of 35.1 seconds."

Ed turned to Ben. "Pridmore here holds the current world record in cards, at 31.03 seconds. And he's British." This elicited a round of rowdy cheers from the guests. "Ben has also learned twenty-seven packs of cards in an hour—which is just, frankly, unnecessary."

Ben unfolded his arms and spoke up. "Lukas and I have been talking, and we've been thinking that since Ed is ranked seventeenth in the world—"

"You mock me," Ed protested. He didn't know that a handful of young Germans had recently passed him in the international rankings.

"We've decided we will not compete unless he can tell us the name of every person in this room."

There were more rowdy cheers, which Ed tried to quiet. He made it about a quarter of the way around the room before getting stumped by a friend of a friend, whom he claimed never to have met. He asked for silence, invited two guests to shuffle the packs of cards, and then handed them to Lukas and Ben. A stopwatch was set. They each had a minute.

Barely a half dozen cards were flipped over before it became clear that Lukas, who had been keeping his head upright only with concerted vigilance, was in no condition to use his higher cognitive faculties. He laid the deck back down on the table and sheepishly announced, "At least I am still ahead of Ed in the international rankings."

Ed forcefully nudged Lukas out of the way and took his seat. "On the occasion of my twenty-fifth birthday, it gives me great pleasure to say that one of the competitors in my showcase event is too drunk to compete and I am going to have to take over!" The decks were reshuffled and the stopwatch reset. "Now, Pridmore, would you calm down, please?"

After a minute of hushed memorization, Ben and Ed took turns announcing cards from memory while a self-appointed judge checked to see that they were correct.

Ed: "Jack of clubs." Cheers.

Ben: "Two of diamonds." Boos.

Ed: "Nine of clubs." Cheers.

Ben: "Four of spades." Boos.

Ed: "Five of spades." Cheers.

Ben: "Ace of spades." Boos.

About forty cards into the deck, Ben shook his head and put his hands down on the table. "That's enough for me."

Ed leaped up from his seat, his breasts slapping his chin. "I knew Ben Pridmore would go too fast! I knew it! He crashes and burns, that guy!"

"How many times have you won the world championship?" Ben responded, with more bite in his voice than I'd ever heard before.

"Shall we clarify our record in one-on-one competition, Ben?"

"You realize losing was my birthday present to you."

As Ed circled the room exchanging high fives and embracing his female guests, Ben slunk back into his bean bag and petted his cape. One of Ed's inebriated Oxford chums, suitably impressed with Ben's performance in spite of his loss, came up to Ben and handed him a short stack of credit cards. He told Ben that if he could memorize them he was welcome to use them.

After the card-off, the party migrated outside to a bonfire that had been built in the clearing, where a drunken tribal hora lasted into the morning. When I finally went to sleep just before sunrise, Ed and Ben were still sitting around the kitchen table, reeling off the most entertainingly bizarre binary number combinations they could think of.

After sleeping off our hangovers, Ed and I spent the next afternoon huddled in training around the kitchen table. I'd come to him with three particular problems I needed his help with, the most pressing of which was that I was consistently mixing up my images. When you're memorizing a deck of cards, there isn't enough time to form images with all the detail and richness that the *Ad Herrennium* calls for. You're moving so fast that usually you can only get the equivalent of a passing glance. In fact, more than anything else, the art of memory is learning how little of an image you need to see to make it memorable. It was only by analyzing the data I was keeping that I realized that I'd been consistently confusing the seven of diamonds—Lance Armstrong riding his bicycle—with the seven of spades—a jockey riding a racehorse. Something about the verb "riding" in those two very different contexts was causing me cognitive hiccups.

I asked Ed what I was supposed to do about that. "Don't try to see the whole image," he said. "You don't need to. Just focus on one salient element of whatever it is you're trying to visualize. If it's your girlfriend, make sure that before all else, you see her smile. Practice studying the whiteness of her teeth, the way her lips crease. The other details will make her more memorable, but the smile will be the key. Sometimes a stab of blue that smells of oysters might be all the recall you get from some particular image, but if you know your system well, you should be able to translate that back again. Often, when you're really gunning for it, the only traces left by a speedily sighted pack of cards will be a series of emotions with no visual content whatsoever. Your other option is to change the images, so they're not so similar—not so mundane."

I closed my eyes and tried to visualize Lance Armstrong pedaling up a steep hill. I made a special point of focusing on the way his reflective sunglasses turned blue and green as they moved through sunlight. Then I thought about the jockey and decided he would be much more distinct as a pony-riding midget with a sombrero. That little adjustment probably shaved two seconds off my time.

"Good stuff with the cards," Ed said when I showed my latest spreadsheet. "It's just a matter of five or so more hours of practice before the images are totally automatic. I've no doubt the American record in speed cards will be child's play. I weep for joy!"

Of course, for all the reanalysis and rejiggering that makes deliberate practice deliberate, Ed warned me that there was always a risk of overthinking things in memory sport, since every change to your mnemonic system leaves behind a trace that can come back to haunt you in competition. And if there's one thing a mental athlete wants desperately to avoid, it's for a single card or number to trigger multiple images on game day.

Another problem I'd discovered in my practice sessions was that my card images were fading too quickly. By the time I'd get to the end of a deck or string of numbers, the images from the beginning had become faint ghosts. I mentioned this to Ed.

"Well, you've got to get to know your images better," was his response. "Starting tonight, take a suit at a time and really spend meditative time with each character. Ask yourself what they look, feel, smell, taste, and sound like; how they walk; the cut of their clothes; their social attitude; their sexual preferences; their propensity to gratuitous violence. After having got this kind of feel for them, try to let it all happen at once—feel the full fat force of their physical and social characteristics all at once in imaginative broadband, and then imagine them going about your house doing everyday things, so you get used to them being so rich and dense even in normal situations. That way, when they do come up in a packet of cards, they should always be offering up some salient characteristic that will stick to their surroundings."

I needed Ed's help with one other problem. Following the recommendations of Peter of Ravenna and the *Ad Herennium*, my collection of PAO images included a handful of titillating acts that are still illegal in a few Southern states, and a handful of others that probably ought to be. And since memorizing a deck of cards with the PAO system requires recombining prememorized images to create novel memorable images, it invariably meant inserting family members into scenes so raunchy I feared I was upgrading my memory at the expense of tormenting my subconscious. The indecent acts my own grandmother has had to commit in the service of my remembering the eight of hearts are truly unspeakable (if not, as I might have previously guessed, unimaginable).

I explained my predicament to Ed. He knew it well. "I eventually had to excise my mother from my deck," he said. "I recommend you do the same."

Ed was a stern coach, who berated me for the "lackadaisical character" of my training. If I went more than a few days without sending him my latest times, or admitted that I was not, in fact, putting in a half an hour a day as he'd commanded, I would receive a caustic reprimand via e-mail.

"You've got to step up your training because it's inevitably the case that your performance will drop in the tournament itself," he warned. "You might have the perfect sporting mentality and actually raise your score, but you've got to work on the assumption that you're going to do better in practice than you'll do in the tournament."

In my own defense, "lackadaisical" wasn't quite the word I'd have chosen. Now that I had put the OK plateau behind me, my scores were improving on an almost daily basis. The sheets of random numbers that I'd memorized were piling up in the drawer of my desk. The dog-eared pages of verse I'd learned by heart were accumulating in my *Norton Anthology of Modern Poetry*. I was beginning to suspect that if I kept improving at the current pace, I might actually have a chance of doing well in the competition.

Ed sent me a quote from the venerable martial artist Bruce Lee, which he hoped would serve as inspiration: "There are no limits. There are plateaus, but you must not stay there, you must go beyond them. If it kills you, it kills you." I copied that thought onto a Post-it note and stuck it on my wall. Then I tore it down and memorized it.

NINE

THE TALENTED TENTH

Not long after returning from England, I found myself sitting on a folding chair in the basement of my parents' home at 6:45 a.m., wearing underpants, earmuffs, and memory goggles, with a printout of eight hundred random digits in my lap and an image in my mind's eye of a lingerie-clad garden gnome (52632) suspended over my grandmother's kitchen table. I suddenly looked up, wondering—remarkably, for the first time—what in the world I was doing with myself.

I realized I'd become fixated on the other competitors. With the help of detailed statistics kept on the memory circuit's stats server, I had made myself familiar with each of their strengths and weaknesses, and I'd measured my own scores against theirs with compulsive regularity. The opponent whom I had become most preoccupied with was not the defending champ, Ram Kolli, a twenty-five-year-old business consultant from Richmond, Virginia, but rather Maurice Stoll, a thirty-year-old beauty-products importer and speed-numbers hotshot from outside Ft. Worth, Texas, who grew up in Germany. I had met him at the previous year's competition. He had a shaved head and a goatee, spoke with an intimidating German accent (anything Germanic is intimidating at a memory contest), and was one of the only Americans to have ever crossed the Atlantic to compete in a European memory contest (he finished fifteenth at the World Memory Championship in 2004 and seventh at that year's Memory World Cup). He held the USA records in both speed numbers (144 digits in five minutes) and speed cards (a single deck in a minute and fifty-six seconds). His only weaknesses were poetry (in which he was ranked ninety-ninth in the world) and insomnia. Everyone agreed he ought to have won the previous year's contest but instead stalled out and finished fourth because he'd only gotten three hours of rest the night before. This year, if he could get to bed on time, I expected he was the favorite to win. And I was now putting in a solid half hour a day to ensure that he didn't.

As I burrowed deeper into my mental training, I was starting to wonder if the sort of memorization practiced by mental athletes was not something like the peacock's tail: impressive not for its utility, but for its profound lack of utility. Were these ancient techniques anything more than "intellectual fossils," as the historian Paulo Rossi once put it, fascinating for what they tell us about the minds of a bygone era, but as out of place in our modern world as quill pens and papyrus scrolls?

That has always been the rap against memory techniques: They're impressive but ultimately useless. The seventeenth-century philosopher Francis Bacon declared, "I make no more estimation of repeating a great number of names or words upon once hearing . . . than I do of the tricks of tumblers, *funambuloes*, *baladines*: the one being the same in the mind that the other is in the body, matters of strangeness without worthiness." He thought the art of memory was fundamentally "barren."

When the sixteenth-century Jesuit missionary Matteo Ricci tried to introduce memory techniques to Chinese Mandarins studying for the imperial civil service exam, he was met with resistance. He planned to hook them first on European study skills before trying to hook them on the European god. The Chinese objected that the method of loci required so much more work than rote repetition, and claimed their way of memorizing was both simpler and faster. I could understand where they were coming from.

The demographics of your average memory contest are pretty much indistinguishable from those of a "Weird Al" Yankovic (five of spades) concert. An overwhelming number of contestants are young, white, male juggling aficionados. Which is why it's impossible to miss the dozen or so students who show up at the USA championship each year in proper church attire. They are from the Samuel Gompers Vocational High School in the South Bronx, and their American history teacher, Raemon Matthews, is a Tony Buzan disciple.

If I had thought that the art of memory was just a form of mental peacocking, Matthews aimed to prove otherwise. He has dubbed the group of students he trains for the USA Memory Championship the "Talented Tenth," after W. E. B. Du Bois's notion that an elite corps of African-Americans would lift the race out of poverty. When I first encountered Matthews at the 2005 USA Memory Championship, he was pacing anxiously at the back of the room, while he waited for his students' scores in the random words event to come in. Several of his students were vying for a top-ten finish, but as far as he was concerned, the real test of their memories was still two and a half months away, when they would sit for the New York State Regents exam. By the end of the year, he expected his students to have memorized every important fact, date, and concept in their U.S. history textbook using the same techniques they employed in the USA Memory Championship. He invited me to come visit his classroom to witness memory techniques being used "in the real world."

To take him up on his offer, I had to pass through a metal detector and have my bag searched by a police officer before entering the Gompers school building. Matthews believes that the art of memory will be his students' ticket out of a neighborhood where nine out of ten students are below average in reading and math, four out of five are living in poverty, and nearly half don't graduate from high school. "The memorization of quotes allows a person to seem more legitimate," he told them, while I sat in the back of his classroom. "Who are you going to be more impressed by, the person who has a litany of his own opinions, or the historian who can draw on the great thinkers who came before him?"

I listened to one student recite verbatim an entire paragraph from *Heart of Darkness* to answer a question about nineteenth-century global commerce. "When it comes time to do the AP test, he'll pull out a quote like that," said Matthews, a dapper dresser with a goatee, closely cropped hair, and a thick Bronx accent. Every in-class essay his students write must contain at least two memorized quotations, just one of many small feats of memory that he demands from them. After school, his students come back for an extracurricular class in memorization techniques.

"It's the difference between only teaching a kid multiplication and giving him a calculator," Matthews says of the memory skills he imparts to them. Not surprisingly, every single member of the Talented Tenth has passed the Regents exam each of the last four years, and 85 percent of them have scored a 90 or better. Matthews has won two citywide Teacher of the Year awards.

Students in the Talented Tenth must wear shirts and ties, and occasionally, at school assemblies, white gloves. Their classroom is plastered with posters of Marcus Garvey and Malcolm X. When they graduate, they receive a kente cloth with the words "Talented Tenth" embossed in gold. At the beginning of each class, the Talented Tenth stand behind their desks, arranged in a pair of facing aisles, and recite in unison a three-minute manifesto from memory that begins: "We are the very best our community has to offer. We will not get lower than 95 percent on any history exam. We are the vanguard of our people. Either walk with our glory and rise to the top with us, or step aside. For when we get to the top, we will reach back and raise you up with us."

The forty-three kids in Matthews's class are all honors students who had to pass a high bar just to get selected for the Talented Tenth. And Matthews works his students hard. "We don't get no vacations," one of them complained to me, while Matthews was standing close enough to overhear. "You work now so you can rest later," he told the student. "You carry your books now so someone else can carry your books later."

The success of Matthews's students raises questions about the purposes of education that are as old as schooling itself, and never seem to go away. What does it mean to be intelligent, and what exactly is it that schools are supposed to be teaching? As the role of memory in the conventional sense has diminished, what should its place be in contemporary pedagogy? Why bother loading up kids' memories with facts if you're ultimately preparing them for a world of externalized memories?

In my own elementary and secondary education, at both public and private schools, I can recall being made to memorize exactly three texts: the Gettysburg Address in third grade, Martin Luther King Jr.'s "I Have a Dream" speech in fourth grade, and Macbeth's "Tomorrow and tomorrow and tomorrow" soliloquy in tenth. That's it. The only activity more antithetical than memorization to the ideals of modern education is corporal punishment.

The slow disappearance of classroom memorization had its philosophical roots in Jean-Jacques Rousseau's polemical 1762 novel, *Émile: Or, On Education*, in which the Swiss philosopher imagined a fictional child raised by means of a "natural education," learning only through self-experience. Rousseau abhorred memorization, as well as just about every other stricture of institutional education. "Reading is the great plague of childhood," he wrote. The traditional curriculum, he believed, was little more than fatuous "heraldry, geography, chronology and language."

The educational ideology that Rousseau rebelled against truly was mind-numbing, and much in need of correction. More than a hundred years after *Émile*'s publication, when the muckraker Dr. Joseph Mayer Rice toured public schools in thirty-six cities, he came away appalled at what he saw, calling one New York City school "the most dehumanizing institution that I have ever laid eyes upon, each child being treated as if he possessed a memory and the faculty of speech, but no individuality, no sensibilities, no soul." At the turn of the twentieth century, rote memorization was still the preferred way to put information, especially history and geography, into kids' heads. Students could be expected to memorize poetry, great speeches, historical dates, times tables, Latin vocabulary, state capitals, the order of American presidents, and much else.

Memorization drills weren't just about transferring information from teacher to student; they were actually thought to have a constructive effect on kids' brains that would benefit them throughout their lives. Rote drills, it was thought, built up the faculty of memory. The *what* that was memorized mattered, but so too did the mere fact that the memory was being exercised. The same was thought to be true of Latin, which at the turn of the twentieth century was taught to nearly half of all American high school students. Educators were convinced that learning the extinct language, with its countless grammatical niceties and difficult conjugations, trained the brain in logical thinking and helped build "mental discipline." Tedium was actually seen as a virtue. And the teachers were backed up by a popular scientific theory known as "faculty psychology," which held that the mind consisted of a handful of specific mental "faculties" that could each individually be trained, like muscles, through rigorous exercise.

Toward the end of the nineteenth century, a group of leading psychologists began to question the empirical basis of "faculty psychology." In his 1890 book *Principles of Psychology,* William James set out to see "whether a certain amount of daily training in learning poetry by heart will shorten the time it takes to learn an entirely different kind of poetry." He spent more than two hours over eight successive days memorizing the first 158 lines of the Victor Hugo poem "Satyr," averaging fifty seconds a line. With that baseline established, James set about memorizing the entire first book of *Paradise Lost*. When he returned to Hugo, he found that his memorization time had actually increased to fifty-seven seconds a line. Practicing memorization had made him worse at it, not better. It was just a single data point, but subsequent studies by the psychologist Edward Thorndike and his colleague Robert S. Woodworth also questioned whether "the general ability to memorize" was influenced by practice memorizing, and found only minor gains. They concluded that the ancillary benefits of "mental discipline" were "mythological" and that general skills, like memorization, were not nearly as transferable as had once been thought. "Pedagogues quickly realized that Thorndike's experiments had undermined the rationale for the traditional curriculum," writes the historian of education Diane Ravitch.

Into this void rushed a group of progressive educators led by the American philosopher John Dewey, who began making the case for a new kind of education that would radically break with the constricted curriculum and methods of the past. They echoed Rousseau's romantic ideals of childhood, and put a new emphasis on "child centered" education. They did away with rote memorization and replaced it with a new kind of "experiential learning." Students would study biology not by memorizing plant anatomy from a textbook but by planting seeds and tending gardens. They'd learn arithmetic not through times tables but through baking recipes. Dewey declared, "I would have a child say not, 'I know,' but 'I have experienced.'"

The last century has been an especially bad one for memory. A hundred years of progressive education reform have discredited memorization as oppressive and stultifying—not only a waste of time, but positively harmful to the developing brain. Schools have deemphasized raw knowledge (most of which gets forgotten anyway), and instead stressed their role in fostering reasoning ability, creativity, and independent thinking.

But is it possible we've been making a huge mistake? The influential critic E. D. Hirsch Jr. complained in 1987: "We cannot assume that young people today know things that were known in the past by almost every literate person in the culture." Hirsch has argued that students are being sent out into the world without the basic level of cultural literacy that is necessary to be a good citizen (what does it say that two thirds of American seventeen-year-olds can't even tell you within fifty years when the Civil War occurred?), and what's needed is a kind of educational counterreformation that reemphasizes hard facts. Hirsch's critics have pointed out that the curriculum he advocates is Dead White Males 101. But if anyone seems qualified to counter that argument it is Matthews, who maintains that for all the Eurocentrism of the curriculum, the fact is that facts still matter. If one of the goals of education is to create inquisitive, knowledgeable people, then you need to give students the most basic signposts that can guide them through a life of learning. And if, as the twelfth-century teacher Hugh of St. Victor put it, "the whole usefulness of education consists only in the memory of it," then you might as well give them the best tools available to commit their education to memory.

"I don't use the word 'memory' in my class because it's a bad word in education," says Matthews. "You make monkeys memorize, whereas education is the ability to retrieve information at will and analyze it. But you can't have higher-level learning—you can't analyze—without retrieving information." And you can't retrieve information without putting the information in there in the first place. The dichotomy between "learning" and "memorizing" is false, Matthews contends. You can't learn without memorizing, and if done right, you can't memorize without learning.

"Memory needs to be taught as a skill in exactly the same way that flexibility and strength and stamina are taught to build up a person's physical health and well being," argues Buzan, who often sounds like an advocate of the old faculty psychology. "Students need to learn how to learn. First you teach them how to learn, then you teach them what to learn.

"The formal education system came out of the military, where the least educated and most educationally deprived people were sent into the army," he says. "In order for them *not* to think, which is what you wanted them to do, they had to obey orders. Military training was extremely regimented and linear. You pounded the information into their brains and made them respond in a Pavlovian manner without thinking. Did it work? Yes. Did they enjoy the experience? No, they didn't. When the industrial revolution came, soldiers were needed on the machines, and so the military approach to education was transferred into school. It worked. But it doesn't work over the long term."

Like many of Buzan's pontifications, this one conceals a kernel of truth beneath an overlay of propaganda. Rote learning—the old "drill and kill" method that education reformers have spent the last century rebelling against—is surely as old as learning itself, but Buzan is right that the art of memory, once at the center of a classical education, had all but disappeared by the nineteenth century.

Buzan's argument that schools have been teaching memory in entirely the wrong way deeply challenges reigning ideas in education, and is often couched in the language of revolution. In fact, though Buzan doesn't seem to see it this way, his ideas are not revolutionary so much as deeply conservative. His goal is to turn the clock back to a time when a good memory still counted for something.

Pinning down Tony Buzan for an interview is no easy task. He is on the road lecturing roughly nine months of the year, and boasts of having racked up enough frequent-flier miles to go to the moon and back eight times. What's more, he seems to cultivate the sense of aloofness and inaccessibility that are a prerequisite for any self-respecting guru. When I finally corralled him behind a desk at the World Memory Championship to discuss the possibility of our sitting down for a couple hours, he opened a large three-ring binder and unfurled a colorful panoramic chart, perhaps three feet long. It was his calendar from the previous year, and it was filled with expansive, continuous blocks of travel—Spain, China, Mexico three times, Australia, America. There was one three-month period when he didn't set foot in the United Kingdom. He told me that he absolutely didn't have any time to speak with me for at least three or four weeks (by which time I would be back home in the United States), but he suggested I visit his estate halfway to Oxford on the river Thames and take some photographs while he was away.

I told him I didn't see how I was likely to learn very much from an empty house.

"Oh, you'd learn quite a lot," he said.

Eventually, through his assistant, I was able to fix an hour with Buzan in his limousine on his way home from the BBC studios in London, where he had just wrapped up a TV interview. I was told to go to a street corner in Whitehall and wait. "You won't be able to miss Mr. Buzan's car."

There was, in fact, no missing it. The car, which pulled up about half an hour late, was a bright ivory 1930s taxicab that looked like it might have just been driven off a BBC set. The door flew open. "Step inside," said Buzan, beckoning. "Welcome to my small, traveling, beautiful lounge."

The first subject we spoke about, because I had to ask, was his unique wardrobe.

"I designed it myself," he told me. He was wearing the same unusual dark navy suit with the large gold buttons that I'd seen him in at the USA championship months earlier. "I used to lecture in an off-the-peg suit, but I was tugging at it with my expansive gestures," he told me. "So I studied fifteenth-, sixteenth-, seventeenth-, eighteenth-, and nineteenth-century swordfighters, and how their arms had not one iota of resistance from their wardrobes. Those ruffles and big sleeves weren't just for show. They were for thrusting and parrying. I design my shirts so that I, too, am free to move."

Everything about Buzan gives the strong impression of someone wanting to make a strong impression. He never swallows a syllable or slouches. His fingernails are as well cared for as the leather of his Italian shoes. There is always a pocket handkerchief tucked neatly in his breast pocket. He signs his letters "*Floreant Dendritae!*"—"May Your Brain Cells Flourish!"—and ends his phone messages "Tony Buzan, over and out!"

When I asked him about the source of his incredible self-confidence, he told me that he owes much of it to his extensive training in the martial arts. He has a black belt in aikido and is three quarters of his way to a black belt in karate. Sitting in the backseat of his limo, he demonstrated a series of jerky moves, a slice through the air, and a shadow punch. "The way I use these techniques is by not using them," he said. "What's the point of fighting if you know you can kill the other, i.e. human, or you can take out his eye, or rip out his tongue?"

Buzan is—he often found occasion to remind me—a modern Renaissance man: a student of dance ("ballroom, modern, jazz"), a composer (influences: "Philip Glass, Beethoven, Elgar"), an author of short stories about animals (under the nom de plume Mowgli, after the boy in *The Jungle Book*), a poet (his last collection, *Concordea*, consists entirely of poems written on and about his thirty-eight transatlantic flights aboard the supersonic Concorde), and a designer (not just of his wardrobe, but also of his home and much of the furniture in it).

About forty-five minutes outside of London, our ivory chariot pulled into Buzan's estate on the river Thames. He asked that I not name its location in print. "Just call it *Wind in the Willows* territory."

Inside his home, named the Gates of Dawn, we took off our shoes and tiptoed around a collection of drawings that had been laid out across the floor, part of an illustrated children's book that he was working on "about a little boy who doesn't do well in school, but does very well in his imagination." There was a large television set with at least a hundred VHS tapes scattered about it, and a bookshelf in the foyer that held the complete *Encyclopaedia Britannica Great Books of the Western World*, several copies of the sci-fi thriller *Dune*, three copies of the Quran, a large quantity of books authored by Buzan, and not much else.

"Is this your library?" I asked.

"I'm only here three months of the year. I have libraries in several other places around the world," he said.

Buzan revels in travel, and in being a man of the world. Once, when I asked him where he's able to find the concentration to turn out two or three books a year, he told me that he has found serene spots to work on almost every continent. "In Australia at the Great Barrier Reef, I write. In Europe, wherever there are oceans, I write. In Mexico, I write. At the Great West Lake in China, I write." Buzan has been traveling since he was a young boy. He was born in London in 1942, but moved with his brother and parents—his mother was a legal stenographer, his father an electrical engineer—to Vancouver at age eleven. He was, he says, "basically a normal kid, in normal trouble, in normal schools."

"My best friend growing up was a boy named Barry," Buzan recalled, sitting outside on his patio with his pink shirt unbuttoned and a pair of large, wraparound geriatric sunglasses protecting his eyes. "He was always in the 1-D classes, while I was in 1-A. One-A was for the bright kids, D for the dunces. But when we went out into nature, Barry could identify things by the way they flew over the horizon. Just from their flight patterns, he could distinguish between a red admiral, a painted thrush, and a blackbird, which are all very similar. So I knew he was a genius. And I got a top mark in an exam on nature, a perfect mark, answering questions like 'Name two fish you can find living in an English stream.' There are a hundred and three. But when I got back my perfect mark on the test, I suddenly realized that the kid sitting down the hall in the dunces' class, my best friend, Barry, knew more than I knew—much more than I knew—in the subject in which I was supposedly number one. And therefore, he was number one, and I was not number one.

"And suddenly, I realized the system that I was in did not know what intelligence was, didn't know how to identify smart and not smart. They called me the best, when I knew I wasn't, and they called him the worst, when he was the best. I mean, there could be no more antipodal environment. So I began to question: What is intelligence? Who says? Who says you're smart? Who says you're not smart? And what do they mean by that?" Those questions, at least according to Buzan's tidy personal narrative, dogged him until he got to college.

Buzan's introduction to the art of memory, the moment that set his entire life on its present path, came, he explained, in the first minutes of his first class on the first day of his first year at the University of British Columbia. His English professor, a dour man "built like a very short wrestler with red tufts of hair on his otherwise bald head" walked into the class and proceeded, with his hands behind his back, to call out the roll of students perfectly. "Whenever someone was absent, he told off their name, their father's name, their mother's name, their date of birth, phone number, and address," recalls Buzan. "And as soon as he'd done it, he looked at us with a sneer on his face. That was the beginning of my love affair with memory."

After class, Buzan charged down the hall after his professor. "I said, 'Professor, how did you do that?' He turned to me and he said, 'Son, I'm a genius.' So I said, 'Sir, that is obvious. But I still want to know how you did it.' He simply said, 'No.' Every day we had English for the next three months, I tested him. I felt he had the Holy Grail, and he wouldn't share it. He despised his students. He thought they were a waste of time. Then one day he said, 'In the beginning of this miserable relationship between myself and yourselves, I demonstrated the exquisite power of human memory and none of you even noticed, so I'm now going to put on the board the code by which I managed to accomplish that extraordinary feat, and I am utterly convinced that none of you will even recognize the treasures put before you—these pearls before swine.' He winked at me and he put up the code. It was the Major System. Suddenly, I realized I could memorize anything."

Buzan left class that day in a trance. It occurred to him, for the first time, that he had not even the most basic idea about how the complicated machinery of his mind worked. And that seemed odd. If the simplest memory trick could dramatically increase the amount of information a person could remember, and nobody had bothered to teach him that trick until he was twenty years old, what else was there that he'd never learned?

"I went to the library and I said, 'I want a book on how to use my brain.' The librarian sent me to the medical section, and I came back and said, 'I don't want a book on how to operate *on* my brain. I want a book on how to operate it. Slightly different.' She said, 'Oh, there are no books on that.' I thought, you get an operations manual on your car, your radio, your TV, but no operations manual on the human brain?" In search of something that might elucidate his professor's feat of memory, Buzan found himself drawn to the library's ancient history section, where his professor had suggested he might find some of the original ideas about improving memory. He began reading up on Greek and Roman mnemonics (in Buzan's pronunciation, the M is not silent), and practicing the techniques in his spare time. It wasn't long before he was using the *Ad Herennium*'s advice about loci and images to study for exams—even to memorize all his notes from entire courses.

After graduating from college, Buzan went on to work a collection of odd jobs in Canada, first as a farmer ("I thought I'd take that job just to have 'shoveling shit' at the top of my CV"), then in construction. In 1966, the same year that Frances Yates published *The Art of Memory*, the first major modern academic work to delve into the rich history of mnemonics, Buzan returned to London to become the editor of *Intelligence*, the international journal of Mensa, the high-IQ society, which he had joined in college. Around the same time, he was hired by the city to work as a substitute teacher at difficult inner-city schools in East London. "I was a special have-brain-will-travel teacher," he says. "If a teacher got beaten up, I was the next one into that classroom."

In most cases, Buzan had just a short amount of time with each of the classes he was subbing for, a few days at most, and hardly enough for even the most well-intentioned teacher to believe he could make any difference. In search of ways to help his troubled students, and perhaps rub off a bit of his own abundant self-confidence on them, Buzan turned to the old memory techniques he had first learned in college. "I would go into the classroom and ask the students whether they were stupid or not, because everyone had been calling them stupid, and sadly they believed they were stupid," says Buzan. "They had been inculcated with the idea of their own incapacity. I said, 'OK, let's check it out,' and I'd give them a memory test, which they'd fail. I'd say, 'Seems you're right about being stupid.' Then I'd teach them a memory technique, and then I'd retest them, and they'd get twenty out of twenty. Then I'd basically say, 'You told me you were stupid, you proved you were stupid, and then you just got a perfect score on a test.' So I'd get them to question: What's going on here? For some of the students who'd never gotten a perfect score on an exam, this was quite a revelation."

Having the opportunity not only to practice the art of memory but also now to teach it allowed Buzan to start developing the old techniques in new directions, particularly when it came to note taking. Over the course of several years, he created what he believed was a completely new system for taking notes that took advantage of the ancient wisdom of the *Ad Herennium.*

"I was trying to get to the essence—the queen's jelly—of what note taking was all about," he says. "That led me to codes and symbols, images and arrows, underlining and color." Buzan called his new system Mind Mapping, a term he later trademarked. One creates a Mind Map by drawing lines off main points to subsidiary points, which branch out further to tertiary points, and so on. Ideas are distilled into as few words as possible and whenever possible are illustrated with images. It's a kind of outline, exploded radially across the page in a rainbow of colors, a web of associations that looks like a prickly bush, or a neuron's branching dendrites. And because it is full of colorful images arranged in order across the page, it functions as a kind of memory palace scrawled on paper.

"In our gross misunderstanding of the function of memory, we thought that memory was operated primarily by rote. In other words, you rammed it in until your head was stuffed with facts. What was not realized is that memory is primarily an imaginative process. In fact, learning, memory, and creativity are the same fundamental process directed with a different focus," says Buzan. "The art and science of memory is about developing the capacity to quickly create images that link disparate ideas. Creativity is the ability to form similar connections between disparate images and to create something new and hurl it into the future so it becomes a poem, or a building, or a dance, or a novel. Creativity is, in a sense, future memory." If the essence of creativity is linking disparate facts and ideas, then the more facility you have making associations, and the more facts and ideas you have at your disposal, the better you'll be at coming up with new ideas. As Buzan likes to point out, Mnemosyne, the goddess of memory, was the mother of the Muses.

The notion that memory and creativity are two sides of the same coin sounds counterintuitive. Remembering and creativity seem like opposite, not complementary, processes. But the idea that they are one and the same is actually quite old, and was once even taken for granted. The Latin root *inventio* is the basis of two words in our modern English vocabulary: inventory and invention. And to a mind trained in the art of memory, those two ideas were closely linked. Invention was a product of inventorying. Where do new ideas come from if not some alchemical blending of old ideas? In order to invent, one first needed a proper inventory, a bank of existing ideas to draw on. Not just an inventory, but an indexed inventory. One needed a way of finding just the right piece of information at just the right moment.

This is what the art of memory was ultimately most useful for. It was not merely a tool for recording but also a tool of invention and composition. "The realization that composing depended on a well-furnished and securely available memory formed the basis of rhetorical education in antiquity," writes Mary Carruthers. Brains were as organized as modern filing cabinets, with important facts, quotations, and ideas stuffed into neat mnemonic cubbyholes, where they would never go missing, and where they could be recombined and strung together on the fly. The goal of training one's memory was to develop the capacity to leap from topic to topic and make new connections between old ideas. "As an art, memory was most importantly associated in the Middles Ages with composition, not simply with retention," argues Carruthers. "Those who practiced the crafts of memory used them—as all crafts are used—to *make* new things: prayers, meditations, sermons, pictures, hymns, stories, and poems."

In 1973, the BBC caught wind of Buzan's work on Mind Mapping and mnemonics and brought him in for a meeting with the network's head of education. The ten-program BBC series and accompanying book that came out of that meeting, both of which were titled *Use Your Head*, helped turn Buzan into a minor British celebrity and made him realize that there was enormous commercial potential in the memory techniques he was promoting. He began taking his ideas, many of which were borrowed directly from the ancient and medieval memory treatises, and repackaging them in a steady stream of self-help books. To date, he's published nearly 120 titles, including *Use Your Perfect Memory*, *Make the Most of Your Mind*, *Use Both Sides of Your Brain*, *Use Your Memory*, and *Master Your Memory*. (At one point, I was alone with Buzan's chauffeur long enough to ask his opinion of his boss's work. "Same meat, different gravy" was his private assessment of Buzan's ouevre.)

To his credit, Buzan is undeniably a marketing genius. He has established franchises of Buzan-licensed instructors all over the world who are trained to teach his memory enhancement, speed reading, and Mind Mapping courses. Today there are over three hundred Buzan-licensed instructors in more than sixty countries. And a thousand teachers around the world are officially teaching Buzan-endorsed memory systems. He estimates that over his entire career the gross sales of all Buzan products, including books, tapes, television programs, training courses, brain games, and lectures, exceeds $300 million.

The competitive memory community breaks cleanly into two camps: those who think Tony Buzan is the second coming of Jesus Christ and those who think he has gotten rich peddling overhyped, sometimes unscientific ideas about the brain. They point out, not unfairly, that while Buzan preaches a "global educational revolution," he has had far more success in creating a global commercial empire than in actually getting his methods into classrooms.

What is especially frustrating for folks like Ed, who take the art of memory seriously and believe in Tony Buzan's basic message that the art of memory still has a place in the modern classroom, is that the messenger can often be a bit of an embarrassment.

Buzan has a troubling habit of lapsing into pseudoscience and hyperbole when he describes how wonderfully revolutionary memory training can be, or how he has "changed the lives of millions." He's been known to say preposterous things, like "Very young children use 98 percent of all thinking tools. By the time they're 12, they use about 75 percent. By the time they're teenagers, they're down to 50 percent, by the time they're in university it's less than 25 percent, and it's less than 15 percent by the time they're in industry."

The fact that Buzan can go around making outrageous claims about the brain and not only be widely believed but actually celebrated is evidence of what a wild frontier the world of brain science is, and how much people want to believe that their memories are improvable. The truth is, the operating manual for the brain that Buzan went looking for in college still hasn't been written.

But for all the pseudoscience and hyperbole that Buzan employs in promoting Mind Mapping, there actually is scientific evidence that his systems work. Researchers at the University of London recently gave a group of students a six-hundred-word passage to read, after teaching half of them how to take notes with a Mind Map. The other half were instructed to take notes normally. When they were tested a week later, the students who used Mind Maps retained about 10 percent more factual knowledge from the passage than the students who used conventional note-taking techniques. That may be a modest gain, but it's hardly insignificant.

My own impression of Mind Mapping, having tried the technique to outline a few parts of this book, is that much of its usefulness comes from the mindfulness necessary to create the map. Unlike standard note-taking, you can't Mind Map on autopilot. My sense is that it's a reasonably efficient way to brainstorm and organize information, but hardly the "ultimate mind-power tool" or "revolutionary system" that Buzan makes it out to be.

Raemon Matthews doesn't have any doubt about the effectiveness of Mind Maps or memory training. At the end of the year, each of his students creates an intricately detailed Mind Map of the entire USA history textbook. Most of the students' maps take up an entire three-panel science-fair board with arrows linking every word and image, from Plymouth Rock in one corner to Monica Lewinsky in the other. "If they get an essay question about the causes of World War I on their AP test, they can just see that part of the map in the mind, and the causes are right there," says Matthews. There might be an image of a black hand to represent the Serbian nationalist organization that Archduke

Franz Ferdinand's assassin belonged to, next to a machine gun wearing running shoes, which represents the arms race that swept Europe in the early years of the twentieth century, and beside that a pair of triangles to represent the Triple Alliance and the Triple Entente.

Matthews takes every opportunity to turn facts into images. "My students were having a hard time getting their heads around the differences in the economic systems of Lenin and Stalin," he told me. "I told them, 'Look, Lenin is sitting on the toilet, and he's constipated because of his mixed economy. Stalin busts into the stall and says, "What are you doing in here?" And Lenin goes, "Land, peace, and bread."' They never forgot that image."

A valid criticism of these sorts of mnemonics is that they are a form of decontextualized knowledge. They are superficial, the epitome of learning without understanding. This is education by PowerPoint, or worse, CliffsNotes. What can an image of Lenin and Stalin in the bathroom really tell you about communist economics? But Matthews's point is that you've got to start somewhere, and you might as well start by installing in students' minds the sorts of memories that are least likely to be forgotten.

When information goes "in one ear and out the other," it's often because it doesn't have anything to stick to. This is something I was personally confronted with not long ago, when I had the opportunity to visit Shanghai for three days while reporting an article. Somehow I had managed to scoot through two decades of schooling without ever learning even the most basic facts about Chinese history. I'd never learned the difference between Ming and Qing, or even that Kublai Khan was actually a real person. I spent my time in Shanghai roving around the city like any good tourist, visiting museums, trying to get a superficial grasp of Chinese history and culture. But my experience of the place was severely impoverished. There was so much I didn't take in, so much I was unable to appreciate, because I didn't have the basic facts to fasten other facts to. It wasn't just that I didn't *know*, it was that I didn't have the ability to *learn*.

This paradox—it takes knowledge to gain knowledge—is captured in a study in which researchers wrote up a detailed description of a half inning of baseball and gave it to a group of baseball fanatics ("experts" is the term Ericsson would use) and a group of less avid fans to read. Afterward they tested how well their subjects could recall the half inning. The baseball fanatics structured their recollections around important game-related events, like runners advancing and runs scoring. They were able to reconstruct the half inning in sharp detail. One almost got the impression they were reading off an internal scorecard. The less avid fans remembered fewer important facts about the game and were more likely to recount superficial details like the weather. Because they lacked a detailed internal representation of the game, they couldn't process the information they were taking in. They didn't know what was important and what was trivial. They couldn't remember what mattered. Without a conceptual framework in which to embed what they were learning, they were effectively amnesics.

Could any less be said of those two thirds of American teens who don't have a clue when the Civil War occurred? Or the 20 percent who don't know who the United States fought against in World War II? Or the 44 percent who think that the subject of *The Scarlet Letter* was either a witch trial or a piece of correspondence? Progressive education reform has accomplished many things. It has made school a lot more pleasant, and a lot more interesting. But it's also brought with it costs for us as individuals and as citizens. Memory is how we transmit virtues and values, and partake of a shared culture.

Of course, the goal of education is not merely to cram a bunch of facts into students' heads; it's to lead them to understand those facts. Nobody would agree with that more than Raemon Matthews. "I want thinkers, not just people who can repeat what I tell them," he says. But even if facts don't by themselves lead to understanding, you can't have understanding without facts. And crucially, the more you know, the easier it is to know more. Memory is like a spiderweb that catches new information. The more it catches, the bigger it grows. And the bigger it grows, the more it catches.

The people whose intellects I most admire always seem to have a fitting anecdote or germane fact at the ready. They're able to reach out across the breadth of their learning and pluck from distant patches. It goes without saying that intelligence is much, much more than mere memory (there are savants who remember much but understand little, just as surely as there are forgetful old professors who remember little but understand much), but memory and intelligence do seem to go hand in hand, like a muscular frame and an athletic disposition. There's a feedback loop between the two. The more tightly any new piece of information can be embedded into the web of information we already know, the more likely it is to be remembered. People who have more associations to hang their memories on are more likely to remember new things, which in turn means they will know more, and be able to learn more. The more we remember, the better we are at processing the world. And the better we are at processing the world, the more we can remember about it.

TEN

THE LITTLE RAIN MAN IN ALL OF US

By February, a month before the USA Memory Championship, my suspicions that I might actually have a chance of doing well in the competition were beginning to be confirmed by my practice scores. In every event except the poem and speed numbers, my best practice scores were approaching the top marks of previous USA champions. Ed told me not to make too much of the fact. "You always do at least twenty percent worse under the lights," he said, repeating advice he'd given me many times before. Still, I was rather stunned by the progress I'd made. In practice, I'd even managed to memorize a deck of cards in one minute and fifty-five seconds, a second faster than the U.S. record. In that day's training log appears this note: "Maybe I could really win this thing?!" (Also, this inscrutable note: "Pay attention to DeVito's remaining hair!!")

What had begun as an exercise in participatory journalism had become an obsession. I had set out simply wanting to learn what the strange world of the memory circuit was all about, and to find out if my memory was indeed improvable. That I might be in a position to really win the USA championship seemed about as improbable as George Plimpton stepping into the ring with Archie Moore and actually knocking him out.

Everything I'd been told—by Ed, by Tony Buzan, by Anders Ericsson—suggested that my course of tedious training was the only way to achieve a more perfect memory. Nobody comes into the world with an inborn ability to remember loads of random digits or poetry at a single glance, or take pictures with the mind.

And yet, combing through the literature, one comes across a few rare cases here and there—perhaps less than a hundred in the last century—of savants with remarkable memories who appear to break the rules. What's most striking about these individuals is that their exceptional memories—"memory without reckoning," it's been called—almost always coexist with profound disability. Some are musical prodigies, like Leslie Lemke, who is blind and brain damaged and couldn't walk until he was fifteen, but can nevertheless play complicated musical pieces on the piano after hearing them just once. Some are artistic prodigies, like Alonzo Clemons, who has an IQ of 40 but can sculpt lifelike animals from memory after just a fleeting glimpse. Some have freakish mechanical skills, like James Henry Pullen, the nineteenth-century "Genius of Earlswood Asylum," who was deaf and nearly mute, but built stunningly intricate model ships.

One day, after memorizing 138 digits in one of my five-minute practice sessions, I was sitting in front of the television, riffling through a deck of cards, as I often did to pass the time. I was looking at the queen of clubs, thinking about Roseanne Barr, about to form a disgusting memory, when I caught a trailer for a new documentary called *Brainman* about one of those rare prodigies. The subject of the film, which aired on the Science Channel, was a twenty-six-year-old British savant named Daniel Tammet, whose brain had been altered by an epileptic seizure he suffered as a toddler. Daniel could perform complex multiplication and division in his head, seemingly effortlessly. He could tell you if any number up to ten thousand was a prime. Most savants have just a single claim to exceptionality, a lone "island of genius," but Daniel had a veritable archipelago. In addition to his lightning calculations, he was also a hyperpolyglot—a term used to describe the small number of people who can speak more than six languages. Daniel claimed to speak ten, and he said he learned Spanish in a single weekend. He'd even invented a language of his own called Mänti. To test his linguistic skills, the producers of *Brainman* flew Daniel to Iceland, and gave him one week to become conversational in Icelandic, one of the world's most notoriously difficult languages. The talk-show host who tested him on national television at the end of the week pronounced himself "amazed." Daniel's tutor for the week called him a "genius" and "not human."

The producers of the *Brainman* documentary also invited two of the world's leading brain scientists, V. S. Ramachandran at the University of California, San Diego, and Simon Baron-Cohen at Cambridge, to each spend a day testing Daniel. They both concluded that he was literally a one-of-a-kind phenomenon. Unlike virtually every other savant who had ever been studied, he could explain what was going on in his head—often in vivid detail. Shai Azoulai, a graduate student in Ramachandran's lab, proclaimed that Daniel "could be the linchpin that spawns off a new field of research." Dr. Darold Treffert, an expert in savant syndrome, declared him one of only fifty people in the world who can be classified as a "prodigious savant."

Even though it's described as a syndrome, savantism is not actually a recognized medical condition, and has no set of standard diagnostic criteria. However, Treffert divides savants into three informal categories. There are "splinter skill" savants who have memorized a single esoteric body of trivia, like Treffert's young patient who can tell you the year and model of a vacuum cleaner just from its unique hum. A second group, which he calls "talented savants," have developed a more general area of expertise, like drawing or music, which is remarkable only because it stands in such stark contrast to their disability. The third group, prodigious savants, have abilities that would be spectacular by any standard, even if they weren't accompanied by handicaps in other areas. It's a subjective scale, but an important one, Treffert believes, because prodigious savants are members of one of the rarest classes of human being on the planet. When a new prodigious savant like Daniel is discovered, it is a very big deal.

The media devoured Daniel's story. Newspapers in England and America ran glowing profiles of the eminently quotable "Boy with the Incredible Brain." He appeared on *The Late Show with David Letterman*, where he calculated the day of the week Dave was born on (Saturday), and on the *Richard & Judy* program, the closest thing Britain has to Oprah. His memoir, *Born on a Blue Day,* became a *New York Times* bestseller in America, and quickly rose to number one in the Amazon UK rankings. Daniel became perhaps the most famous living savant in the world.

What interested me most about Daniel was his extraordinary memory. In 2003, he set a new European record by reciting the first 22,514 digits of pi from memory. It took him five hours and nine minutes, sitting in the basement of the Science Museum at Oxford University, and he says he did it without any mnemonic techniques beyond his powerful raw memory. Here, it seemed, was someone with the same astounding abilities as the mental athletes, but they came to him entirely without effort. It was almost impossible to believe. Meanwhile, I was putting in torturous hours taking mental strolls through every home I'd ever visited, every school I'd ever attended, and every library I'd ever worked in so that they could be converted into memory palaces. I wondered why a savant like Daniel never competed in memory contests. Surely he'd wipe the floor with the trained mnemonists, I imagined.

The more I researched Daniel's story, the more fascinated I was by the differences between him and the mental athletes I'd come to know—and the mental athlete I was rapidly becoming myself. I knew how the mnemonists did it: They'd improved their memories through rigorous training, using ancient techniques. I'd even done it myself. But I didn't understand where Daniel's powers of recall came from. Daniel, like the journalist S before him, seemed to have an innate ability to remember. How was his brain different from mine? And did he have any tricks up his sleeve that could give me an advantage at the USA championship?

I decided that I would try to meet up with Daniel. He invited me to the home he shared with his partner, Neil, at the end of a leafy cul-de-sac in the scenic seaside town of Kent, England. We ended up spending two full afternoons together in his living room, chatting over tea and fish and chips. Daniel was skinny, with short blond hair, glasses, and birdlike features. He was gentle, soft-spoken, charming, and hyperarticulate—equally comfortable explaining his bizarre memory as opining on why *The West Wing* was the most thoughtful American television program. I suppose I'd come expecting some kind of freak, and so I was taken aback by how surprisingly ordinary Daniel seemed—even more ordinary than some of the mental athletes I'd come to know. In fact, if he hadn't told me, I'm not sure I'd ever have guessed that there was anything unusual about him. However, Daniel assured me that despite appearances, he was anything but normal. "You should have met me fifteen years ago. You'd have said, 'Boy, that guy has autism!' "

Daniel is the oldest of nine children. He grew up in subsidized housing in East London and had what he calls "a very difficult" childhood that "seems like something out of Dickens." In *Born on a Blue Day*, he describes the massive epileptic seizure he suffered as a four-year-old: It was "an experience unlike any other, as though the room around me was pulling away from me on all sides and the light inside it leaking out and the flow of time itself coagulated and stretched out into a single lingering moment." Had his father not rushed him to the emergency room in the back of a taxi, that seizure very probably would have killed Daniel. Instead, he believes it was the moment he became a savant.

According to Baron-Cohen, two rare conditions may have conspired to produce Daniel's savant abilities. The first is synesthesia, the same perceptual disorder that afflicted the journalist S, in which the senses are intertwined. By one estimate, there are more than a hundred different varieties of the disorder. For S, sounds conjured up visual imagery. In Daniel's case, numbers take on a distinctive shape, color, texture, and emotional "tone." The number 9, for example, is tall, dark blue, and ominous, while 37 is "lumpy like porridge" and 89 resembles falling snow. Daniel says he has a unique synesthetic reaction like that for every number up to 10,000, and that experiencing numbers in this way allows him to do quick mental math without pencil or paper. To multiply two numbers, he sees each number's shape floating in his mind's eye. Intuitively, and without effort, he says, a third shape, the answer, forms in the negative space between them. "It's like a crystallization. It's like developing a photo," Daniel told me. "Division is just the reverse of multiplication. I see the number and I pull it apart in my head. It's like leaves falling from a tree." Daniel believes his synesthetic shapes somehow implicitly encode important information about the properties of numbers. Prime numbers, for example, have a "pebble-like quality." They're soft and round, without the jagged edges of numbers that can be factorized.

Daniel's other rare condition is Asperger's syndrome, a form of high-functioning autism. Autism was first identified in 1943 by the child psychiatrist Leo Kanner. He described it as a form of social impairment, a disorder in which, as Kanner put it, patients "treat people as if they were things." Along with this inability to empathize, autistic individuals have a host of other problems, including language impairment, an extremely focused range of interests, and "an anxiously obsessive desire for the preservation of sameness." A year after Kanner first wrote about autism, an Austrian pediatrician named Hans Asperger noted another disorder that seemed almost identical except that Asperger's patients had strong linguistic abilities and fewer intellectual impairments. He called his precocious young patients, with their bottomless wells of arcane trivia, "little professors." It wasn't until 1981 that Asperger's was recognized as its own separate syndrome.

Daniel's Asperger's diagnosis was made by Baron-Cohen, who runs the Cambridge Autism Research Centre and who also happens to be one of the world's leading authorities on synesthesia. "If you saw him today, you wouldn't necessarily think that this guy has a form of autism," Baron-Cohen told me over tea in his Trinity College office one afternoon. "It's only in the context of hearing his developmental history. I said to him, 'Your development suggests that when you were younger you had Asperger's syndrome, whereas looking at you today, you've made such a good adaptation and you're functioning so very well that you don't necessarily need a diagnosis. It's up to you whether you want one or not. He said, 'Yes, I prefer to have it.' It gave him a new way of looking at himself. That's fine. It fits with his profile."

In his memoir, Daniel writes extensively about the effects of growing up with undiagnosed Asperger's. "What must the other children have made of me? I don't know, because I have no memory of them at all. To me they were the background to my visual and tactile experiences." Throughout his childhood, Daniel was afflicted with a passion for trivia. He collected leaflets and counted everything, and developed an obsessive, encyclopedic knowledge of the popular 1970s soft-rock duo the Carpenters. He frequently got into trouble for taking things far too literally. After extending his middle finger in the direction of a schoolmate, he was surprised at the reprimand he received. "How can a finger swear?" he wondered. Empathy did not come easily. "I had no concept of deception," he says. "I've worked so hard to reach this place where I can be really normal, where I can have a conversation and know when to start and stop talking, and remember to make eye contact." Despite having apparently conquered his most debilitating social problems, to this day, Daniel says he still can't shave himself, or drive a car. The sound of the toothbrush scratching his teeth drives him mad. He says he avoids public places, and is obsessive about small things. For breakfast, he measures out exactly forty-five grams of porridge on an electric scale.

I mentioned *Brainman* to Ben Pridmore. I was curious to know whether he'd seen it, and whether he was afraid that Daniel, someone with natural gifts that seemed to measure up to—if not surpass—Ben's own acquired skills, might someday make an appearance on the memory circuit.

"I'm pretty sure that guy *did* compete in the championships a couple years ago," Ben told me matter-of-factly. "But I think he had a different name. Back then he was called Daniel Corney. He did quite well one year, as I recall."

I asked a few of the other mental athletes what they thought of Daniel. Almost everyone had seen *Brainman*, and almost everyone had an opinion. Quite a few were suspicious about his claims of savanthood, and believed he used basic mnemonic techniques to memorize information. "Any of us could do what he's done," the eight-time world memory champion Dominic O'Brien told me. "If you want my opinion, he simply realized he'd never be number one as a mental athlete." O'Brien said as much on camera, when he was filmed for *Brainman*, but the producers didn't include his interview in the final cut.

Clearly the mental athletes had plenty of reason to be envious of Daniel. His memory skills were almost exactly equivalent to theirs, and yet their respective places in the cultural firmament couldn't have been more different. While the trained mnemonists toiled away in geeky obscurity, Daniel's medicalized condition had generated enormous popular interest.

The next time I was in front of a computer, I logged into the memory circuit stats server. Sure enough, I found a Daniel Corney who had competed twice in the World Memory Championship, finishing as high as fourth place in 2000. It was the same Daniel, with a different surname: He'd had it legally changed in 2001. It seemed strange to me that in his memoir about his impressive memory Daniel wouldn't have mentioned his fourth-place finish in the World Memory Championship.

I did a search for Daniel's name in the Worldwide Brain Club, the online forum where mental athletes gather. Not only had Daniel competed in the World Memory Championship, he had actually been an outspoken critic of it, even going so far as to lay out an eight-point program for how memory sport could be made more legitimate, more popular, and attract more media attention. I was especially surprised by one of Daniel's posts to the WWBC. It was an ad from the year 2001 in which he offered to reveal the "secrets of his 'Mindpower formula' in his unique 'Mindpower and Advanced Memory skills e-mail course.'" What secrets were those? I wondered. And why hadn't he shared them with me when we met?

What fascinates us and excites us about savants—the reason Daniel has received so much attention from both scientists and the public—is their otherness, and their ability to do the seemingly impossible with apparent ease. They are, in effect, aliens in our midst, walking exceptions to the natural order of the universe. As jaw-dropping as the memory tricks performed by mental athletes may be, they're still just tricks. And like any magic trick, once you know how it's done—and that you could do it, too—the effect loses a good bit of its luster. But savants are the real deal: For them, memory is not a trick, but a talent.

But now I was beginning to wonder if the gulf between me and Daniel—between any of us and Daniel—might not be nearly so great at it seemed. What if, as Dominic O'Brien seemed to believe, the most famous savant in the world was not a rare individual with almost mystical natural abilities but just a guy who accomplished savantlike tricks through methodical training? What, then, would be the difference between him and me?

When it comes to savant memory, there is probably only one other human being in the same class as Brainman: Kim Peek, aka Rain Man, the prodigious savant born in 1951 who inspired Dustin Hoffman's character in the Hollywood movie. He has arguably the best memory in the world. Now that I'd spent some time with Daniel, I decided to visit Kim in his hometown in Utah to make a comparison, to find out what the two celebrated prodigies had in common, and what they could tell me about the nature of savant syndrome.

I met Kim on a leg of what has become his endless speaking tour—on which his father and caregiver, Fran, accompanies him, and for which he never requests payment. He was addressing a group of about three dozen elderly women in the activities room at an old-age home in his hometown of Salt Lake City. Members of the audience had been invited to try to stump him with obscure trivia (anything but "logic or reasoning questions," Fran cautioned). A woman breathing from an oxygen tank asked him about the highest peak in South America.

He answered correctly—Mt. Aconcagua, a fact any mildly competent trivia buff would know—and gave its height: 22,320 feet (which, I later discovered, was off by about five hundred feet). An amputee in a wheelchair inquired how many times Easter fell in March in the 1930s. Without a pause, he responded. "March 27, 1932. March 28, 1937." His answers ended with a quickening of his voice that sounded like it was about to explode in raucous laughter. The program director of the home asked him which books were summarized in volume 4 of *Reader's Digest Condensed Books* from 1964. He named all five. The name of Harry Truman's daughter? Margaret. The number of times the Steelers have won the Super Bowl? Four. The last line of Coriolanus? "Which to this hour bewail the injury, / Yet he shall have a noble memory. Assist."

"He's never forgotten anything," Fran told me, including, supposedly, every fact in the more than nine thousand books he has read at about ten seconds a page. (Each eye scans its own page independently.) He's memorized Shakespeare's entire corpus and the scores to every major piece of classical music. At a recent staging of *Twelfth Night*, an actor transposed two lines, sending Kim into a fit of such magnitude that the house lights had to be turned on and the play suspended. He's no longer allowed to attend live plays.

Unlike Daniel, there's no way to look at Kim and not immediately sense that he is entirely unique. He has gray hair and a bearlike build, and squints through thick, brown plastic frames. His head is almost always tilted forty-five degrees to the side. He keeps one hand docked inside the other, and thrusts it in and out when he gets excited. Possibly the most allusive conversationalist on the planet, his mind so overflows with facts and figures that they often come out as a waterfall of apparent non sequiturs. When an Argentine woman at the old-age home told Kim that she was born in Córdoba, Kim immediately told her the major roads into and out of her hometown and then belted out the chorus of "Don't Cry for Me, Argentina," provoking a squirm of discomfort from me. And then out of nowhere he screamed, "You're fired!" Fran helped him explain the connection: The basketball star Dennis Rodman, who used to date Madonna, who played Argentinean first lady Eva Perón in the movie version of *Evita*, was fired by the Los Angeles Lakers in 1999.

Kim seems to have discovered a Pavlovian association between his astounding literalness and audience laughter. At a recent talk, he responded to a question about the content of the Gettysburg Address with, "227 Northwest Front Street. But Lincoln stayed there only one night. He gave the speech the next day." He now repeats that joke often.

Kim likes to be called the "Kimputer," but his full name is Laurence Kim Peek. "We named him after Laurence Olivier and Rudyard Kipling's Kim," says Fran. When Kim was born, after a difficult pregnancy, it was immediately clear that something was deeply off. His head was a third larger than normal and sprouted a fist-size blister on its backside that the doctors were afraid to remove. For the first three years of his life, Kim dragged his head on the ground as if it were loaded with a heavy weight. He didn't walk until he was four. His parents were urged to consider a lobotomy. Instead Kim was put on heavy sedatives until he was fourteen. Fran recalls that it was only when Kim was taken off the sedatives that he first started to show an interest in books. He's been memorizing them ever since.

But even though Kim has access to a larger store of knowledge than perhaps anyone else on the planet, he doesn't seem able to put it toward any end other than itself. He has an IQ of just 87. And no matter how many books of etiquette he may have memorized, his sense of what's socially appropriate is, to put it generously, esoteric. Standing in a crowd of people in the lobby of the Salt Lake City public library, Kim wrapped his thick arms around my shoulders and gripped me against his paunch and then forcibly gyrated against me. "Joshua Foer, you are a great, great man," he told me loudly enough to startle a passerby. "You are a handsome man. You are a man of your generation." And then he let out a deep roar.

How Kim can do what he does is a mystery to science. Unlike Dustin Hoffman's character in *Rain Man*, Kim is not, apparently, autistic. He's far too sociable for that diagnosis. He's something else entirely. In January 1989, the same week that *Rain Man* was released, a CT scan of Kim's brain revealed that his cerebellum, an organ crucial to sensory perception and motor function, was severely distended. An earlier scan had discovered that Kim also lacks a corpus callosum, the thick bundle of neurons that connects the left and right hemispheres of the brain, and allows them to communicate. It's an incredibly rare condition, but how it might contribute to his savantism isn't at all clear.

Kim and I spent the better part of our afternoon together sitting at a table in the back corner of the Salt Lake City public library's fourth floor, where he has spent almost every weekday of the last ten years reading and memorizing phone books. He took off his glasses and laid them on the table. "I'm just going to do a little scanning," he announced. I looked over his shoulder as he leafed through a phone book from Bellingham, Washington. I was trying to keep pace with his memory. I did what Ed would have coached me to do had he been there: I set up a memory palace and converted each person's phone number into an image, did the same thing with the first and last name, and then quickly tried to tie all those images together in a memorable way. It was hard work, and when I tried to explain it to Kim, he didn't seem to understand what I was talking about. Every time I'd get to the fourth or fifth name in the first column, he was ready to move on to the next page. I asked him how he was able to do it so quickly. He looked up from the book and peered over his glasses, agitated by my interruption. "I just remember!" he screamed. And then he reburied his head in a column of phone numbers, and ignored me for the next half hour.

One of the challenges of developing a theory to explain savant syndrome is that it expresses itself so differently in different individuals. However, there is one neuroanatomical anomaly that turns up again and again in savants, including Kim: damage in the brain's left hemisphere. Interestingly, the exaggerated abilities of savants are almost always in right-brain sorts of activities, like visual and spatial skills, and savants almost always have trouble with tasks that are supposed to be primarily the left-brain's domain, such as language. Speech defects are extremely common among savants, which is part of the reason that loquacious, well-spoken Daniel seems so extraordinary.

Some researchers have theorized that shutting off certain left-brain activities somehow liberates right-brain skills that had been latent all along. Indeed, people have been known to suddenly acquire savantlike abilities later in life, after a traumatic injury to the left side of the brain. In 1979, a ten-year-old boy named Orlando Serrell took a baseball pitch to the left side of his head and came to with a remarkable capacity to calculate calendar dates and remember the weather on every day of his life. Bruce Miller, a neurologist at the University of California, San Francisco, studies elderly patients with a relatively common form of brain disease called frontotemporal dementia, or FTD. He's found that in some cases where the FTD is localized on the left side of the brain, people who had never picked up a paintbrush or an instrument can develop extraordinary artistic and musical abilities at the very end of their lives. As their other cognitive skills fade away, they become narrow savants.

The fact that people can become savants so spontaneously suggests that those exceptional abilities must lie dormant, to some degree, in all of us. There may be, as Treffert likes to put it, "a little Rain Man" hiding inside every brain. He's just kept under lock and key by the inhibitory "tyranny of the dominant left hemisphere."

Treffert further speculates that savants with exceptional memories may somehow hand over the duties of maintaining declarative memories, like facts and figures, to the more primitive nondeclarative memory systems, like those that help us recall how to ride a bike or catch a fly ball without consciously thinking about it (the same systems that allow the amnesic HM to draw in the mirror and EP to navigate his neighborhood without knowing his address). Consider how much mental processing must take place just to position one's hand to catch a fly ball—the instantaneous calculations of distance, trajectory, and velocity—or to recognize the difference between a cat and a dog. Our brains are obviously capable of astoundingly fast and complex calculations that happen subconsciously. We can't explain them because most of the time we hardly even realize they're happening.

But with enough effort those lower levels of cognition can sometimes be accessed. For example, when students are taught to draw, often the first two exercises they're made to master are tracing negative space and contour lines. The aim of these exercises is to shut down the top-level conscious processing that can't see a chair as anything but a chair, and activate the latent, lower-level perceptual processing that sees it only as a collection of abstract shapes and lines. It takes a great deal of training for an artist to learn to deactivate that top-level processing; Treffert believes savants may do it naturally.

If the rest of us could turn off that top-level processing, would we become savants? There actually is a technology that can selectively, and temporarily, turn off parts of the brain. It's called transcranial magnetic stimulation, or TMS, and it works by using focused magnetic fields to wreak havoc on the electrical firing of targeted neurons. The deadening effect can last for upwards of an hour. Although TMS is relatively new, it has been used successfully as a noninvasive means of treating problems as diverse as depression, post-traumatic stress disorder, and migraines. But in many ways, TMS's experimental potential is even more exciting than its therapeutic uses. There are obviously some intractable ethical problems with experimenting on the human brain.

Since you can't go in and mess around with a living brain (HM taught us that), much of what neuroscientists have been able to learn about the brain has been the result of a few "natural experiments" caused by extremely unlikely forms of brain damage (like EP's). Because TMS allows neuroscientists to turn regions of the brain on and off at will, they can use it to perform repeatable experiments without waiting for someone to walk into their office with a rare lesion that just happens to affect the specific part of the brain they want to study. Allan Snyder, an Australian neuroscientist who popularized TMS as an experimental tool, uses the technique to temporarily induce savantlike artistic skills in otherwise normal people by targeting the left frontotemporal lobe (the same region that is often damaged in savants). After having the left temporal lobe zapped, subjects can draw more accurate pictures from memory, and can more quickly estimate the number of dots flashed on a screen. Snyder calls his device a "creativity-amplifying machine." He might as well call it the savant cap.

In the *Brainman* documentary, I had watched Daniel divide 13 by 97 and give the result to so many decimal places that the answer ran off the edge of a scientific calculator. A computer had to be brought in for verification. He multiplied three-digit numbers in his head in a few seconds, and quickly figured out that 37 to the fourth power was 1,874,161. To me, Daniel's mental math seemed much more impressive than his memory.

As I began looking into the complicated subject of mental math, I learned that just like mnemonics, the field has its own vast literature, and even its own world championship. Indeed, with a bit of Googling and a whole lot of practice, anyone can teach themselves how to multiply three-digit numbers in their head. It is by no means easy—believe me, I tried—but it's a skill that can be learned.

Though I asked him repeatedly on several occasions, Daniel refused to perform even a single mental calculation for me. "One of my parents' big fears was that I would become a freak show," he said when I pressed him. "I had to promise them that I wouldn't do calculations for people who ask me. I only do them for scientists." But he did perform some mental math for the cameras in *Brainman*.

As he was performing those calculations, I was struck by something odd that Daniel seemed to be doing with his fingers. While he's supposedly watching the answer crystallize in his mind's eye, the camera captures his index finger sliding around on the surface of the desk in front of him. Given his descriptions of shapes melting and fusing in his mind, that little bit of finger work just struck me as strange.

Talking to a few experts, I learned that anyone who has done mental multiplication might have suspicions about those sliding fingers. One of the most common techniques for calculating the product of two large numbers is known as cross multiplication. It involves doing a sequence of individual multiplications of single-digit numbers and then combining them together in the end. To my eye, this appeared to be what Daniel may have been doing on the table. Daniel denies this. He says it's just a fidget that helps him concentrate.

"There are a lot of people in the world who can do those kinds of things, but they're still pretty impressive," Ben Pridmore told me. In addition to competing on the memory circuit, Ben also competes in the Mental Calculation World Cup, a biennial contest in which participants carry out mental calculations far more extreme than Daniel's, including multiplying eight-digit numbers without pencil or paper. None of these top calculators make any claims about seeing numerical shapes that fuse and divide in their minds' eyes. They all readily admit to using techniques detailed in countless books and Web sites. I asked Ronald Doerfler, author of one of those books, *Dead Reckoning: Calculating Without Instruments,* to watch *Brainman* and tell me what he thought. "I'm not fantastically impressed with any of that," he said of Daniel's mathematical talents, and added, "The lore of mental calculators is rife with misdirection."

What about the fact that Daniel knows all the prime numbers less than 10,000? Even this doesn't impress Ben Pridmore. "Just basic memorization," he says. There are only 1,229 primes less than 10,000. That's a lot of numbers to commit to memory, but not compared to learning 22,000 digits of pi.

Calendar calculating, the only savant skill Daniel was willing to perform in front of me, turns out to be so simple that it really shouldn't impress anyone. Savants like Kim, who can tell you the date of every Easter in the last thousand years, seem to have internalized the rhythms and rules of the calendar without explicitly understanding them. But anyone can learn them. There are several very simple calendar calculation formulas, published widely on the Internet. It only takes about an hour of practice to become fluent with them.

The more Daniel and I talked, the more his own statements began to cast doubt on his story. When I asked him on different occasions two weeks apart to describe what the number 9,412 looked like, he gave me two completely different answers. The first time he said, "There's blue in there because it starts with a nine, and a drifting motion as well, and kind of like a sloping as well." Two weeks later, he said after a long pause, "It's a spotty number. There's spots and curves as well. It's actually a very complex number." Then he added, "The larger the numbers are, the harder they are to put into words. That's why in interviews, I usually concentrate on the smaller numbers." Indeed, synesthetes are never entirely consistent, and to his credit, Daniel described several smaller numbers consistently over the course of our meetings.

But what about those "Mindpower and Advanced Memory skills" courses that Daniel used to advertise on the WWBC? Back at his home in Kent, I handed him a printout of his ad from 2001 and asked him what I was supposed to make of it. If his extraordinary memory came to him entirely without effort and he didn't need to use mnemonic techniques, why was he selling a course on exactly that subject? He uncurled his feet from under him and put them back on the floor. "Look, I was twenty-two at the time," he said. "I had no money. The one thing I had experience in was competing in the World Memory Championship. So I wrote a course on improving your memory. When I went to the world championship, I found out that the people taught themselves to remember. None of them had good memories. I thought at the time that they were lying, but it did give me the idea that this was something you could teach. I was in a position where I had to sell myself. The only thing I thought was sellable was my brain. So I used Tony Buzan kind of stuff. I said, 'Expand your brain,' and that sort of thing, but I didn't like doing it."

"You don't use memory techniques?" I asked him.

"No," he assured me.

If Daniel had concocted his story of being a natural savant, it would have required a degree of mendacity that I couldn't quite bring myself to believe he possessed. If he was merely a trained mnemonist trying to cloak himself in the garb of a savant, why would he so willingly subject himself to scientific testing?

How could one ever know whether Daniel is what he claims to be? For a long time, scientists were skeptical that synesthesia even existed. They dismissed the phenomenon as fakery, or nothing more than lasting associations made between numbers and colors in childhood. Despite all the case reports in the literature, there was no way of proving that something so seemingly far-fetched was actually taking place in someone's brain. In 1987, Baron-Cohen developed the Test of Genuineness for Synesthesia, the first rigorous assessment of the condition. The test measures the consistency with which a purported synesthete reports color-word associations over time. When Baron-Cohen administered a version of this test to Daniel, he passed it with ease. Still, I couldn't help but wonder if any trained mnemonist would have been able to do as well. Other results from Daniel's scientific testing struck me as odd. When Baron-Cohen tested Daniel's memory for faces, he performed abysmally, leading Baron-Cohen to conclude that "his face memory appears impaired." That sounds like just the sort of thing a savant might be bad at. And yet when Daniel Corney competed in the World Memory Championship, he won the gold medal in the names-and-faces event. It just didn't make sense.

One test that might help demonstrate Daniel's synesthesia more conclusively would be an fMRI scan. In many number-color synesthetes, you can actually see brain areas associated with color processing light up when the subject is asked to read a number. When Baron-Cohen teamed up with fMRI experts to look at Daniel's brain, they didn't find this. Their test subject "did not activate extra-striate regions normally associated with synaesthesia suggesting that he has an unusual and more abstract and conceptual form of synaesthesia," the researchers concluded. Were it not for the fact that he'd passed the Test of Genuineness for Synesthesia, another reasonable conclusion might be that Daniel is not a synesthete at all.

"Sometimes people ask me if I mind being a guinea pig for the scientists. I have no problem with it because I know that I am helping them to understand the human brain better, which is something that will benefit everyone," Daniel writes in his memoir. "It is also gratifying for me to learn more about myself, and the way in which my mind works." When Anders Ericsson invited Daniel to visit FSU to be tested according to his own exacting standards, Daniel said he was too busy to make the trip.

The problem with all the tests given to Daniel is their null hypothesis—the working assumption that would be true if their alternative hypothesis were proven false: namely, that if Daniel wasn't a savant, then he must be just a regular guy. But what needs to be tested, especially in light of his unusual personal history, is the alternative possibility that the world's most famous savant might actually be a trained mnemonist.

About a year after my first meeting with Daniel, his publicist e-mailed me to ask if I wanted to meet with him again, this time over breakfast at the stylish midtown hotel he was staying at in New York. He was in town to do an appearance on *Good Morning America* and to promote his book, *Born on a Blue Day*, which had debuted in America in third place on the *New York Times* nonfiction bestseller list.

After a cup of coffee and some pleasant chitchat about his life in the spotlight, I asked him again—for the third time—what the number 9,412 looked like to him. There was a flicker of recognition in his eyes before he closed them. He knew I hadn't pulled those digits out of thin air. He put his fingers in his ears, and held them there for two very long, uncomfortable minutes of silence. "I can see it in my head. But I can't break it down," he said, finally.

"Last time I asked, you were able to describe it almost immediately."

He thought about it a bit longer. "It would be dark blue, and it would be pointy, and shiny, with a drifting motion. Or I could picture it as ninety-four and twelve, in which case it would look like a triangle and this sort of shape." He made a kind of quadrilateral with his arms. His face was cherry red. "It depends on all sorts of things, like whether I heard the number OK, and how I decided to break it up. It depends on whether I'm tired. I make mistakes sometimes. I see the wrong number. I mistake it for a number that looks similar. That's why I prefer to do tests with actual scientists. There isn't the same stress."

I read back to him the descriptions he'd given me of 9,412 the last two times I'd seen him. They could hardly have been more different. I told him my theory, which I realized would be very difficult to prove: that he was using the same basic techniques as other mental athletes, and that he invented these far-out synesthetic descriptions of numbers to mask the fact that he had memorized a simple image to associate with each of the two-digit combinations from 00 to 99—one of the most basic techniques in the mnemonist's tool kit. It was one of the most uncomfortable sentences I'd ever spoken to anyone.

For some time, I agonized over whether to include Daniel in this book. But late one night, not long before I was supposed to turn in a draft of this chapter, I decided to do one last Internet search for his name—just to see if I might have missed something, or at least to refresh my memory about a story that had been sitting in a folder in my filing cabinet for over a year. Somehow—and I don't know how I missed this before—I found a cached version of danieltammet.com, a Web site created in 2000 that hadn't been online for at least three or four years. The seven-year-old "About" page describing Daniel included a surprisingly forthright bit of autobiography that didn't make it into *Born on a Blue Day*:

> My own interest in memory and conversely memory sport was sparked by my casual acquaintance with a children's book on broad memory concepts for better exam performance at the age of 15. The following year I passed my GCSEs with some of the year's best results and subsequently performed well at A-level, mastering French and German along the way with the help of these tried-and-tested techniques . . . My obsession with the sport grew, and following months of strenuous training and hard work I climbed into the World's Top-5 rated Memory sportsmen.

Earlier, I had also found something else, a series of messages posted several years ago from the same e-mail address used by Daniel Corney, but sent by someone named Daniel Andersson, who claimed to be "a well-respected and gifted psychic with more than 20 years of experience helping and empowering others." The messages explained that Daniel Andersson had received his psychic powers during a series of childhood seizures. There was a link to a Web site where you could arrange a phone reading with Daniel for "advice on all manner of topics, including relationship problems, health and financial matters, lost love and contacting those who have passed over."

I asked Daniel what I was supposed to make of those e-mails. Six years ago he was claiming that his epileptic seizures gave him psychic powers. Now he was claiming that they'd made him a savant. "Do you see why someone might be suspicious?" I asked.

He paused to collect himself. "God this is embarrassing," he said. "After I offered myself as a tutor and that wasn't successful, I read an ad for someone who could do psychic readings. You could work from home and use the telephone. That was ideal for me. I wasn't a psychic. I did it for about a year because I had no income otherwise. I was regularly told off, because I wasn't giving advice. I was mostly just listening. I treated it, start to finish, only as an opportunity to listen to people. With hindsight, I wish I hadn't done that work. But I was desperate. Look, life is complex. I never thought I would come into the public eye. I promise you that I've done tests for consistency with scientists who are well placed to determine whether I'm putting it on, and their opinion—not just one scientist's opinion, crucially—is that I'm for real."

Toward the end of our final meeting, I told Daniel all the reasons I couldn't bring myself to believe that he, the world's most famous savant, was truly a savant. "I want to be convinced," I told him, "but I'm not."

"If I wanted to trick you, if I wanted to pull the wool over your eyes, I would practice immensely," he said frankly. "I would come out all guns blazing. I would jump through every hoop. But I sincerely don't care what you think about me. I don't mean that in a personal way. I mean that I don't care what anybody thinks about me. I know myself. I know what goes on in my head when I close my eyes. I know what numbers mean to me. These things are hard to explain, and hard to put into terms you can easily analyze. If I was some very good person at defending something, then I would think very carefully and make some great impression on you and everyone else."

"You *have* made a great impression on everyone else."

"People trust scientists and scientists have studied me—and I trust scientists. They're neutral. They're not media. They're not interested in writing a particular angle. They're interested in truth. With media, I am just who I am. Sometimes I'll come across very well, other times I will be more nervous, and I won't make such a good impression. I'm human. I'm inconsistent because I'm human. Of all the people who've interviewed me, you have treated me the most like a normal person. You've not idolized me. You've treated me on your level. I respect that. I feel more comfortable being a human than being an angel."

"That may be because I suspect you are just a normal person," I said. As those words came out of my mouth, I realized I didn't really mean them. What frustrated me about Daniel was that I knew he *wasn't* normal. In fact, the one thing I know I can say for certain about him is that he is exceptionally bright. I know how much work it takes to train one's memory. Anyone can do it, but not just anyone can do it to the degree that I suspected Daniel had. I believed Daniel was special. I just wasn't sure he was special in the way he was claiming.

I asked Daniel if, when he looked at himself in the mirror honestly, he really considered himself a savant.

"Am I a savant?" He put down his coffee and leaned in close. "It all depends on how you define the word, doesn't it? You could define 'savant' in such a way that I would be excluded from the definition. You could define it such a way that Kim Peek would be excluded from the definition. And you could define it in such a way that there would be no more savants in the world at all."

It all comes down to definitions. In his book *Extraordinary People,* Treffert defines savant syndrome as "an exceedingly rare condition in which persons with serious mental handicaps...have spectacular islands of ability or brilliance which stand in stark, markedly incongruous contrast to the handicap." According to that definition, the question of whether Daniel uses memory techniques would be irrelevant to whether he is a savant. All that matters is that he has a history of a developmental disability and can perform phenomenal mental feats. According to Treffert's definition, Daniel would indeed be a prodigious savant, albeit one whose disability is less pronounced. However, what Treffert's definition does not capture is the clear difference between someone like Kim Peek, whose incredible abilities are apparently unconscious and perhaps even automatic, and someone who achieves those same skills through tedious, methodical training.

As late as the nineteenth century, the term "savant" had an entirely different connotation than it has today. It was the highest epithet that could be bestowed on a man of learning. A savant was someone who had mastered multiple fields, who traded in abstract ideas, who "consecrate[d] their energies to the search for truth," as Charles Richet, the author of the 1927 book *The Natural History of a Savant,* put it. The term had nothing to do with singular abilities or a prodigious memory. And yet over the last century the word's meaning has changed. In 1887, John Langdon Down, better known for the chromosomal disorder that bears his name, coined the term "idiot savant." The word "idiot," regarded as politically incorrect, eventually fell away. In a world in which our everyday memories have atrophied and we've become totally estranged from the idea of a disciplined memory, "savant" has gone from being a term of art and an emblem of intellectual accomplishment to being a freakish condition, a syndrome. You'd never hear a polymath like Oliver Sacks described as a savant today, though he, as much as anyone, meets the dictionary definition. Today, the word is reserved for people like the autistic twins that Sacks famously wrote about, who were supposedly able to count 111 matches the instant they spilled onto the floor.

So what about someone like Daniel? One of the oldest myths about savants is that they were destined to be born into this world as geniuses, but by some terrible twist of fate had all of their aptitudes curtailed but one. I wonder about Daniel. I wonder what we would say about him if he was just a guy who had trained himself to memorize 22,000 digits of pi and to multiply three-digit numbers in his head. I wonder what we'd say if he'd achieved those things only through rigorous discipline and enormous effort. Would that make him more incredible than Kim Peek, or less? We want to believe that there are Daniel Tammets walking among us, individuals born into this world with extraordinary talents in the face of extraordinary difficulties. It is one of the most inspiring ideas about the human mind. But perhaps Daniel exemplifies an even more inspiring idea: that we all have remarkable capacities asleep inside of us. If only we bothered ourselves to awaken them.

ELEVEN

THE USA MEMORY CHAMPIONSHIP

There was to be a new event at the 2006 USA Memory Championship, one never before experienced in the history of memory competitions. It was clunkily called "Three Strikes and You're Out of the Tea Party," and it had been dreamed up specifically to please the producers from HDNet, the cable network that would, for the first time ever, be airing the contest on national television. Five strangers, posing as guests at a tea party, would walk onto the stage and tell the competitors ten pieces of information about themselves—their addresses, phone numbers, hobbies, birthdays, favorite foods, pets' names, the make and model of their cars, etc. It was a test as true to the demands of real life as there had ever been in a memory contest. I had no idea how I would prepare for it, and frankly I hadn't thought much about it until just a month and a half before the contest, when Ed and I spent a pair of evenings on a transatlantic telephone call inventing a system that would allow me to quickly and easily file away all of that personal information in a specially designed memory palace set aside for each of the strangers.

I had constructed five imaginary buildings, one for each of the "tea party" guests. Each was built in a different style, but with a similar floor plan based around a central atrium and satellite rooms. The first palace was a modernist glass cube in the manner of Philip Johnson's Glass House; the second was a turreted Queen Anne of the type you see all over San Francisco, with lots of frilly scrollwork and ostentatious ornamentation; the third was Frank Gehry–esque, with wavy titanium walls and warped windows; the fourth was based on Thomas Jefferson's redbrick home, Monticello; and there was nothing special about the fifth except that all the walls were painted bright blue. Each home's kitchen would serve as the repository of an address. Each home's den would hold a phone number. The master bedroom was for hobbies, the bathroom was for birthdays, and so on.

Three weeks before the contest, after reviewing the scores I'd been sending him, Ed phoned to tell me that I needed to stop practicing all other events and begin focusing exclusively on the tea party. I rounded up friends and family and had them make up fictional biographies for me to memorize in my painstakingly appointed new palaces. Several unromantic dinners with my girlfriend were spent with her in character, telling me stories about her life as a Nebraska farmer or a suburban housewife or a Parisian seamstress, which I then recalled for her over dessert.

Then, one week before the championship, just at the moment when I wanted to be training hardest, Ed told me I had to stop. Mental athletes always halt their training a week before contests in order to do a spring cleaning of their memory palaces. They walk through them and mentally empty them of any lingering images, because in the heat of competition, the last thing you want to do is accidentally remember something you memorized last week. "Some competitors, when they get to a really high level, will not speak to anyone three days before a contest," Tony Buzan told me. "They feel that any association that enters their head could interfere with associations they form in the contest."

The plan had always been for Ed to be ringside at the USA championship. But shortly before the contest, he shipped off to Australia, where he'd been offered a unique opportunity to do philosophy research at the University of Sydney on the phenomenological issues raised by the sport of cricket. (He believes that the sport contains even better examples than chicken sexers or chess grand masters to illuminate his thesis that our immediate perception of the world is powerfully shaped by memory.) Suddenly it was no longer certain that he would be able to make the much longer and more expensive trip from the other side of the earth.

"Is there any way I can mediate your disgust at my potential nonappearance?" he asked in an e-mail a couple days before the contest. The emotion I was feeling was not so much disgust as panic. Though I'd told everyone I knew that I was approaching the contest as little more than a whim—"a strange way to spend a weekend morning" was how I put it to a friend—the jokes I sometimes made at the expense of this "kooky contest" concealed the truth that I was dead set on victory.

Ed's last-minute decision to stay in Australia meant that I was on my own to worry about the other competitors, to speculate on how intensely they'd trained over the last year, and to wonder whether any of them were preparing to surprise us by unveiling a new technique that would take the sport to a level I could not reach. There was Ram Kolli, the cheery and insouciant defending champ, who I knew was the most natural talent of the group. If he had decided to train as hard as a European, the rest of us wouldn't have a chance. But somehow I doubted he had it in him. Mostly I fretted about Maurice Stoll. If anyone might have committed the time to developing a Millennium PAO system like Ed's, or a 2,704-image card system like Ben's, I suspected it would be Maurice.

The evening before the championship, Ed e-mailed me one last piece of advice: "All you have to do is to savor the images, and really enjoy them. So long as you're surprising yourself with their lively goodness, you'll do just fine. Don't at any stage worry. Take it easy, ignore the opposition, have fun. I'm proud of you already. And remember, girls dig scars and glory lasts forever."

That night, I lay in bed obsessively marching through each of my palaces—first forward and then backward—and worrying about Maurice. I couldn't sleep, which, as Maurice himself had observed at the previous year's competition, is for a mental athlete "like breaking your leg before a soccer match."

When I finally did get to sleep sometime around three a.m., with the assistance of some Tylenol PM, I had a terrifying dream in which Danny DeVito and Rhea Perlman, my king and queen of spades, were riding around a parking garage for hours on a pony, the seven of spades, trying in vain to find where they'd parked their Lamborghini Countach, the jack of hearts. Eventually they and their horse melted into the asphalt, while Maurice Stoll looked on with a sinister Dr. Mengele cackle. I got up four hours later, bleary and dazed, and accidentally shampooed my hair twice—an ominous portent if ever there was one.

The first person I ran into when I got off the elevator on the nineteenth floor of the Con Edison headquarters was Ben Pridmore. He had flown in from England for the weekend solely to scout out the American field. At the airport in Manchester, he had splurged on a last-minute first-class upgrade. "What else have I got to waste my money on?" he asked me. I looked down at his half-eaten leather shoes, whose soles were now almost entirely detached. "Good point," I said.

"The first event hasn't even started, and I've already lost," I told Ben. I explained about my insomnia and my redundant shampooing, and he seemed convinced that I had done myself no favors with those sleeping pills, whose chemicals, he said, were probably still swimming around in my bloodstream.

I downed two tall cups of coffee and, in truth, felt more jittery than tired. Mostly I just felt stupid for having so catastrophically screwed up the most important thing I needed to do in order to be competitive. Meanwhile, Maurice walked in wearing a Texas Longhorns baseball cap and a paisley shirt, looking far perkier than he had last year. And frighteningly confident. He recognized me from across the room, and strode straight over to shake my hand and introduce himself to the legendary Ben Pridmore.

"You're back," Maurice said to me. It was an assertion, not a question. To the extent that I had a strategy, it was to sneak up on Maurice and surprise him. But apparently he'd already been briefed on me. Somebody must have informed him that I'd been training with Ed Cooke.

"Yeah, I thought I'd try competing this year," I said nonchalantly, and pointed down at my name tag, which read "Joshua Foer, Mental Athlete." "It's kind of like a journalistic experiment."

I asked, "How are your numbers looking this year?" I was probing him to see if he'd upgraded his system.

"They're good. And yours?"

"Good. What about cards?"

"Not bad. You?"

"I should be all right in cards," I said. "Still using the same systems as last year?"

He shrugged a nonreply and asked me, "How did you sleep last night?"

"What?"

"How did you sleep?"

Why was he asking me that? How did he know about my insomnia? What kind of head games was Maurice trying to play? "Remember, last year I didn't sleep so good," he continued.

"Yeah, I remember that. And this year?"

"This year, I slept just fine."

"Josh needed sleeping pills," said Ben helpfully.

"Yeah, well, they're basically a placebo, right?"

"I tried to take sleeping pills one time in practice, and I fell asleep the next morning memorizing numbers," said Maurice. "You know, lack of sleep is the enemy of memory."

"Oh."

"Anyway, good luck today."

"Yeah, good luck to you, too."

New this year was the gaggle of TV cameras buzzing about the room and the play-by-play analysts—the mixed martial arts announcer Kenny Rice and his color man, the four-time U.S. champ Scott Hagwood—perched in front of the stage on director's chairs. Their presence lent the contest the surreal quality of a mockumentary. Did I really just hear Rice describe the contestants as having "taken mental prowess to a whole new level"?

Unlike the international competitions I'd been to, where competitors spent the moments before a contest isolated between a pair of earmuffs or juggling to warm up their brains, the USA competitors all milled about making small talk, as if they were about to take a test no more demanding than an eye exam. I sequestered myself in a corner, inserted my earplugs, and tried to clear my mind like a proper European mnemonist.

Tony Dottino, a slim, silver-haired, and mustachioed fifty-eight-year-old corporate management consultant, stood at the front of the room to introduce the contest. Dottino founded the USA Memory Championship in 1997 and has run thirteen of them ever since. He is one of Tony Buzan's American disciples. Dottino makes his living consulting with companies like IBM, British Airways, and Con Edison (hence the unlikely location of the championship) about how their workforces can be made more productive through the use of memory techniques.

"You are the folks telling people in our country that memory is not for geeks," he declared. "You will be the models that people will come to follow. We're like little infants in terms of writing the history of these events. You"—he pointed at us with both index fingers—"are writing the history books." I tuned out for the rest of his speech, put my earplugs back in, and took one last walk through each of my palaces. I was checking, as Ed had once taught me, to make sure all of the windows were open and good afternoon sunlight was streaming in, so that my images would be as clear as possible.

Among those of us who would contribute to "writing the history books" were three dozen mental athletes from ten states, including a Lutheran minister from Wisconsin named T. Michael Harty, about a half dozen kids from Raemon Matthews's Talented Tenth, and a forty-seven-year-old professional memory trainer from Richmond, Virginia, named Paul Mellor, who had run a marathon in each of the fifty states and had been in New Jersey the previous week teaching police officers how to quickly memorize license plate numbers.

The big guns were all put behind desks in the back row. These were the folks that Dottino had predicted might make a run at the title. I was flattered to be counted among them, albeit in the last seat at the end of the row. (Dottino and I had spoken several times over the previous year, and I'd kept him updated on my practice scores, so he knew I had a sporting chance.) The lineup included a compact thirty-year-old software engineer from San Francisco named Chester Santos, who goes by the nom de guerre "Ice Man," which hardly befits his soft-spoken, aw-shucks manner. He'd finished in third place the previous year. I had a strong suspicion that Chester didn't like me very much. After I'd written my original article for *Slate* about the previous year's USA championship, I was forwarded an e-mail he'd penned to Tony Dottino. In it Chester complained that my piece was "HORRIBLE" because I had made Lukas and Ed "sound awesome," while the USA competitors came off as "complete amateurs and slackers." That I now had the impudence to go head-to-head with him after just a year's training must have seemed like the ultimate insult.

From the sidelines, I heard Kenny Rice say, "It must be intimidating, much like the weekend athlete who wants to take on LeBron James in a game of one-on-one." I figured he was talking about me.

Though every other national memory championship in the world is sewn together from approximately the same standard set of events, according to the same standard set of rules established by the World Memory Sport Council, the United States does things slightly differently. In the international events, everyone's scores are added up at the end of the tournament to determine the winner, but the U.S. championship is less straight forward. It consists of a preliminary morning round of four classic pen-and-paper disciplines—names and faces, speed numbers, speed cards, and the poem—that are used to select six finalists. Those six finalists then compete in the afternoon in three unique television-friendly "elimination" events called "Words to Remember," "Three Strikes and You're out of the Tea Party," and "Double Deck'r Bust," which whittle the field down until there is only one United States memory champion left standing.

The first event of the morning was names and faces, which I'd always done pretty well with in practice. The point of the game is to take a packet of ninety-nine head shots and memorize the first and last name associated with each of them. One does that by dreaming up an unforgettable image that links the face to the name. Take, for example, Edward Bedford, one of the ninety-nine names that we had to remember. He was a black man with a goatee, a receding hairline, tinted sunglasses, and an earring in his left ear. To connect that face to that name, I tried to visualize Edward Bedford lying on the bed of a Ford truck, then, deciding that wasn't distinctive enough, I saw him fording a river on a floating bed. To remember that his first name was Edward, I put Edward Scissorhands on the bed with him, shredding the mattress as he paddled it across the river.

I used a different trick to remember Sean Kirk, a white guy with a mullet, sideburns, and the cockeyed smile of a stroke victim. I paired him up with the Fox News anchor Sean Hannity and Captain Kirk of the Starship Enterprise, and painted an image in my mind of the three of them forming a human pyramid.

After fifteen minutes of the contestants staring at those names and faces, a judge came by and picked up our packets, and handed us a new bunch of stapled pages, with the same set of faces arranged in a different order, and this time, with no names attached. We had fifteen minutes to recall as many of them as possible.

When I put down my pen and handed in my recall sheet, I assumed my score was going to be somewhere near the middle of the pack. The names of Sean Kirk and Edward Bedford had come right back to me, but I'd blanked on the cute blonde and the toddler with the French-sounding name, and a handful of others, so it was hard to imagine I'd done all that well. To my surprise, the 107 first and last names I was able to recall were good enough for a third place finish, just behind Ram Kolli, who memorized 115, and just ahead of Maurice Stoll, who did 104. The winner of the event was a seventeen-year-old competitive swimmer from Mechanicsburg, Pennsylvania, named Erin Hope Luley, who'd managed an impressive 124 names, a new USA record and a score that would have gotten a nod of respect even from the top

Europeans. When her number was announced, she stood up and waved sheepishly. I looked over at Ram, and caught him looking back at me. He lifted his eyebrows as if to say, "Where'd she come from?"

The second event of the morning was speed numbers, always my worst. This was the one event where Ed's coaching had given me little advantage—because I had largely ignored Ed's coaching. He had been prodding me for months to develop a more complicated system for numbers—not quite the "64-gun Man of War" Millennium PAO system he had spent months working on, but something at least a step ahead of the simple Major System that most of the other Americans would be using. I'd indulged him and developed a PAO system for all fifty-two playing cards, but I never got around to doing the same for every two-digit combination from 00 to 99.

Employing the same Major System as the rest of the mental athletes, I used my five minutes of memorization time to go for what I figured was a very safe ninety-four digits—mediocre even by American standards. And still I managed to get the eighty-eighth digit mixed up (instead of Bill Cosby, I should have seen a family playing an oversize version of Milton Bradley's Game of Life). I blamed my poor showing on Maurice, whom I had heard even through my earmuffs gruffly yelling, "Enough with the pictures already!" at a press photographer who was circulating in the room. Still, my eighty-seven digits left me in fifth place. Maurice had banked 148, a new U.S. record, and Ram had finished in second with 124. Erin was way down in eleventh place, having remembered just fifty-two digits. I got up, stretched, and had a third cup of coffee. "They're known as MAs, or mental athletes," I heard Kenny Rice earnestly tell the camera, "but at this point in the competition, MA could stand for something else: mental anguish."

Though I'd been operating with inferior mnemonotechnics in the numbers event, when it came to speed cards, the next challenge, I was the only competitor armed with what Ed referred to as "the latest

European weaponry." Most of the Americans were still placing a single card in each locus, and even the guys who'd been competing for years, like Ram and "Ice Man" Chester, were at best turning two cards into a single image. In fact, only a couple of years ago it was entirely unheard of for anyone to be able to memorize a whole pack of cards at the USA championship. Thanks to Ed, the PAO system I was using packed three cards into a single image, which meant that it was at least 50 percent more efficient than what was being used by any of the other Americans. It was a huge advantage. Even if Maurice, Chester, and Ram were going to wipe me in the other disciplines, I hoped I might be able to run up my score in speed cards.

Each competitor was assigned an individual judge with a stopwatch, who took a seat across the table. Mine was a middle-aged woman, who smiled as she sat down and said something that I couldn't make out through my earplugs and earmuffs. I had brought along my black spray-painted memory goggles for speed cards, and up until the moment a freshly shuffled deck was placed on the desk in front of me, I was still weighing whether to put them on. I hadn't practiced without my goggles in weeks, and the Con Edison auditorium was certainly full of distractions. But there were also three television cameras circulating in the room. As one of them zoomed in for a close-up of my face, I thought of all the people I knew who might end up watching the broadcast: high school classmates I hadn't seen in years, friends who had no idea about my memory obsession, my girlfriend's parents. What would they think if they turned on their TVs and saw me wearing huge black safety goggles and earmuffs, thumbing through a deck of playing cards? In the end, my fear of public embarrassment trumped my competitive instincts, and I left the goggles on the floor by my feet.

From the front of the room, the chief arbiter, a former marine drill sergeant, shouted, "Go!" My judge clicked her stopwatch, and I began peeling through the pack as fast as I could, flicking three cards at a time off the top of the deck and into my right hand. I was storing the images in the memory palace I knew better than any other, the house in Washington, D.C., that I'd lived in since I was four years old—the same house I'd used to remember Ed's to-do list on the rock in Central Park. At the front door, I saw my friend Liz vivisecting a pig (two of hearts, two of diamonds, three of hearts). Just inside, the Incredible Hulk rode a stationary bike while a pair of oversize, loopy earrings weighed down his earlobes (three of clubs, seven of diamonds, jack of spades). Next to the mirror at the bottom of the stairs, Terry Bradshaw balanced on a wheelchair (seven of hearts, nine of diamonds, eight of hearts), and just behind him, a midget jockey in a sombrero parachuted from an airplane with an umbrella (seven of spades, eight of diamonds, four of clubs). Halfway through the deck, Maurice's Teutonic wail once again penetrated my earmuffs: "No walking!" I heard him yell, presumably at another photographer. This time, I didn't let it break my focus. In my brother's bedroom, I saw my friend Ben urinating on Benedict XVI's papal skullcap (ten of diamonds, two of clubs, six of diamonds), Jerry Seinfeld sprawled out bleeding on the hood of a Lamborghini in the hallway (five of hearts, ace of diamonds, jack of hearts), and at the foot of my parents' bedroom door, myself moonwalking with Einstein (four of spades, king of hearts, three of diamonds).

The art of speed cards is in finding the perfect balance between moving quickly and forming detailed images. You want to catch just enough of a glimpse of your images so as to be able to reconstruct them later, without wasting precious time conjuring up any more color than necessary. When I put my palms back down on the table to stop the clock, I knew that I'd hit a sweet spot in that balance. But I didn't yet know how sweet.

The judge, who was sitting opposite me, flashed me the time on her stopwatch: one minute and forty seconds. Not only was that better than anything I'd ever done in practice, but I also immediately recognized that it would shatter the old United States record of one minute and fifty-five seconds. I closed my eyes, put my head down on the table, whispered an expletive to myself, and took a second to dwell on the fact that I had possibly just done something—however geeky, however trivial—better than it had ever been done by anyone in the entire United States of America.

I looked up and quickly glanced over at Maurice Stoll, who was stroking his goatee and seemed agitated, and I felt an unseemly satisfaction in the trouble he seemed to be having. Then I looked over at Chester and got nervous. He was smirking confidently. He shouldn't have been. He had clocked in at a lethargic two minutes and fifteen seconds.

By the standards of the international memory circuit, where thirty seconds is the best time, my minute and forty seconds would have been considered middling—the equivalent of a five-minute mile for any of the serious Europeans. But we weren't in Europe.

As word of my time traveled across the room, cameras and spectators began to assemble around my desk. The judge pulled out a second unshuffled deck of playing cards and pushed them across the table to me. My task now was to rearrange the unshuffled pack to match the one I'd just memorized.

I fanned the unshuffled deck out across the table, took a deep breath, and walked through my palace one more time. I could see all the images perched exactly where I'd left them, except for two. They should have been in the shower, dripping wet, but all I could spy were blank beige tiles.

I can't see it, I whispered to myself frantically. *I can't see it*. I ran through every single one of my images as fast as I could. Had I forgotten a giant pair of toes? A fop wearing an ascot? Pamela Anderson's rack? The Lucky Charms leprechaun? An army of turbaned Sikhs? No, no, no, no.

I began sliding the cards I did remember around with my index finger. In the top left corner of the desk, I put my friend Liz and her dead pig. Next to her, the Hulk on his bike, and Terry Bradshaw with his wheelchair. As the clock ran out on my five minutes of recall time, I was left with three cards still on the table. They were the three cards that had disappeared from the shower: the king of diamonds, four of hearts, and seven of clubs. Bill Clinton copulating with a basketball. How could I have possibly missed it?

I quickly neatened up the stack of cards into a square pile, shoved them back across the table to the judge, and removed my earmuffs and ear plugs. I had it nailed. There wasn't a doubt in my mind.

After waiting a moment for one of the television cameras to circle around for a better angle, the judge began flipping the cards over one by one, while, for dramatic effect, I did the same with the deck I'd memorized.

Two of hearts.

Two of hearts.

Two of diamonds.

Two of diamonds.

Three of hearts.

Three of hearts . . .

Card by card, each one matched. When we got to the end of the decks, I threw the last card down on the table, and looked up with a wide, stupid grin that I tried and failed to squelch. I was the new USA record holder in speed cards. The throng that had gathered around my desk applauded loudly. One person hooted. Ben Pridmore pumped his fist. A twelve-year-old boy stepped forward, handed me a pen, and asked for my autograph.

For reasons that were never made clear, it had been decided that the three top finishers in the first three events of the morning would be given a bye, and wouldn't have to compete in the final preliminary event of the morning, the poem. Despite my low score in numbers, my record performance with the cards was enough to leave me in second place overall, behind Maurice and ahead of "Ice Man" Chester. We were all going straight to the quarterfinals. The three of us left the competition hall with Ben Pridmore and walked over to the Con Edison cafeteria, where we sat at the same table eating a cordial, and mostly silent, lunch. When we returned, the three of us were joined on the stage by Ram, the forty-seven-year-old fifty-state marathoner Paul Mellor, and seventeen-year-old Erin Luley, who had set a new United States record—her second of the day—in the poetry event, while we were out of the room.

Now that there were only six of us left, the competition shifted to its second phase, designed to amp up the drama for the benefit of the television cameras. Nifty 3-D graphics were now projected onto a screen in the front of the room, and theatrical lighting poured down on the stage, where there were six tall director's chairs for us to sit on, each with a lapel microphone resting on it.

The first event of the afternoon was random words. In a typical random words event at a typical national championship, the competitors would have fifteen minutes to memorize as many words as possible from a list of four hundred, then a short break, and then thirty minutes to write as many as they could remember in order on a sheet of paper. It's not exactly a spectator sport. For the U.S. championship, it was decided that everything would happen on stage, with the hope that this might lend the event some of the hand-wringing, agonizing screams, and other kabuki antics that make the spelling bee such compelling theater. The six of us were to go in a circle, one by one, each calling out the next word on the list we'd memorized. The first two mental athletes to miss a word would be knocked out.

The list was a collection of concrete nouns and verbs like "reptile" and "drown," which are the easiest to visualize, mixed in with a few harder-to-imagine abstract words like "pity" and "grace." Whereas your objective in a normal random words event would be to memorize as much as possible, and perhaps be a little reckless about it for the sake of packing your memory palace to capacity, Ed and I had reckoned that the rules of the USA championships meant that a wiser strategy was to memorize fewer words—I went for a mere 120—but make sure they were 100 percent right. We figured most of the people on the stage could probably remember more words than me, but also that somebody was going to freak out and try for more than he or she could handle. I would not be that guy.

After our fifteen minutes of memorizing, we went person by person across the stage announcing the next word from the list: "sarcasm" . . . "icon" . . . "awning" . . . "lasso" . . . "torment" . . . When we got to the twenty-seventh word, Erin, who had just that morning memorized more poetry than any American mental athlete ever before, floundered. The word was "numb"—the other five of us all knew it—but for some reason she couldn't see it. She collapsed back into her chair, shaking her head. Nine words later, Paul Mellor mistook "operation" for "operate"—a classic rookie error. Most of us—and especially the producer from HDNet, which was televising the scintillating proceedings—had been braced for a bruising battle of attrition past at least the hundredth word. It was hard to figure how the event could have ended so early. Even someone who has just learned the principle of the memory palace is usually able to memorize at least thirty or forty words on a first attempt. I suspected that both Erin and Paul had misjudged the rest of the field and overreached. Which meant Ram, Chester, Maurice, and I had slid into the final four on the unforced errors of others. Which meant I was one tea party away from the finals of the USA Memory Championship.

. . .

A tall brunette in a summer dress walked onto the stage and introduced herself. "Hi, I'm Diana Marie Anderson. I was born on December 22, 1967, in Ithaca, New York, 14850. My work number, but please don't call me there, is 929-244-6735, extension 14. I have a pet and her name is Karma and she's a yellow lab. I have some hobbies: watching movies, cycling, and knitting. My favorite car is a 1927 Model T Ford. It's black. When I eat, I have pizza and jelly beans and peppermint-stick ice cream."

While she spoke, Ram, Chester, Maurice, and I had our eyes closed, furiously painting images in our memory palaces. Diana's birthday, 12/22/67, became a one-ton weight (12) crushing a nun (22) as she drank a fruit shake (67), which I placed in a freestanding claw-toothed bathtub in the bathroom of my Victorian palace. For her birthplace and zip code, I walked over to the linen closet and imagined a monster truck tire (14) rolling over the ledge of one of Ithaca's famous gorges, and landing on a couple of fellas (850). Four more tea party guests appeared on stage, and read off equally exhaustive biographies.

The contest was called "Three Strikes and You're Out," which meant that the first two contestants to forget three pieces of information would be eliminated. After giving us a few minutes for the curve of forgetting to work its magic, the five tea party guests came back onstage and started quizzing us about themselves. First, we were asked for the name of a young woman with blond hair and a baseball cap, the fourth of the five guests. Chester, sitting at the end of the row, knew it: "Susan Lana Jones." Maurice was then quizzed on her date of birth, which he didn't know, and which made me wonder if he hadn't been bluffing about his good night's sleep. One strike for Maurice. Fortunately, I did know her birthday. I pulled it out of the stark marble sink of my modernist palace. It was December 10, 1975. Ram knew her place of residence: North Miami Beach, Florida, 33180, but Chester couldn't remember her phone number. One strike for Chester. And neither could Maurice. Two strikes for Maurice. The camera zoomed in on me, waiting for me to call out the ten digits, plus extension. "I didn't even try to remember her phone number," I said, looking straight into the lens. My strategy had been to focus on everything else, and just hope that those long numbers would be someone else's problem. One strike for Josh.

The game continued like this, until it got back to Maurice, who couldn't come up with even a single one of the woman's three hobbies. In fact, he might as well have been taking a nap while they'd been reading off their bios. He had three strikes. He was out.

The other three of us remained on stage volleying biographical details back and forth for several more rounds. Eventually it came back to Chester to recite the work phone number of one of the tea party guests, including the area code and three digit extension.

Chester grimaced and looked down. "Why do I always get the phone numbers? Are you kidding me?"

"That's just the way it worked out," said Tony Dottino, who was standing behind a podium at stage left, acting as game show host.

"Come on, nobody knows the phone numbers."

"You're a numbers guru, Chester."

If I'd been sitting in Chester's chair, I wouldn't have known it either. It was dumb luck that Chester had ended up in that seat and not me, dumb luck that he got his third strike before me, and dumb luck that I was now on my way to the finals of the USA Memory Championship.

A ten-minute pause was announced before the final event, "Double Deck'r Bust," in which Ram and I would each have five minutes to memorize the same two decks of playing cards. Maurice grabbed me as I walked off the stage and put his arm around my shoulder. "You are the winner," he said in clipped English. "Ram cannot do two decks. It is certain." I thanked him curtly, and tried to make my way through the crowd to get out of the room. Ben greeted me at the bottom of the stairs with an outstretched palm waiting for a low five.

"Cards are Ram's worst event," he said excitedly. "You've got it in the bag now!"

"Come on, man, what are you trying to do, jinx this?"

"All you've got to do is half of what you did this morning."

"Please don't say that. You're bringing down some serious evil eye over here."

He apologized and left to find Ram to offer him his best wishes.

From the sideline, Kenny Rice continued his play-by-play analysis: "We are nearing the deciding moment here in the USA National Memory Championship. Ram Kolli won this event last year. Can the twenty-five-year old from Virginia pull off the repeat, or will it be the newcomer Joshua Foer, an Internet journalist who has covered the event before? Now he's here trying to win it. This last event, 'Double Deck'r Bust,' is a mind-against-mind battle."

I knew, despite the bad karma, that Ben and Maurice were right. Ram could barely memorize a single deck of cards in five minutes, much less two. Under the sweat-inducing lights, eye to eye with the lens of a television camera, I knew that all I had to do was not choke, and that silver hand with the golden nail polish would be mine.

The first thing I did after sitting down and putting in my ear plugs was shove the second deck aside. Since I only needed to memorize one more card than Ram, I decided I would get to know the first deck as thoroughly as I possibly could. I spent the five minutes looking at those fifty-two cards over and over again, breaking only to take a quick peek at Ram, who was sitting at the table next to me. He was holding up a single card and studying it like some sort of rare insect. *Oh my god, that guy doesn't have a chance*, I thought.

After five minutes of memorizing, there was a coin toss to determine who would go first during recall. Ram called tails. It was heads. It was up to me to choose whether to start, or let Ram.

"This is important," I whispered, loud enough to be picked up by my lapel microphone. I closed my eyes and walked as fast as I could through the deck, checking to see if there were any gaps in my memory palace, places where for some reason an image hadn't stuck, as had happened earlier that morning. If there were, I wanted Ram to be accountable for those cards, not me. Finally, after a long pause, I opened my eyes. "I'll start."

I thought about it a second longer. "No, no, no. Wait. Ram can start." It might have seemed like one last little bit of psychological gamesmanship, but in fact I'd realized I couldn't remember the forty-third card in the deck. I wanted to make sure that that one would be Ram's responsibility.

Dottino: "Okay, Ram, it's to you for the first card."

Ram twiddled his fingers for a second. "Two of diamonds."

Then me: "Queen of hearts."

"Nine of clubs."

"King of hearts."

Ram looked up at the ceiling and leaned back in his chair.

I could see he was shaking his head. *No freaking way*, I thought. He looked back down. "King of diamonds?"

Now I was shaking my head. I knew he was out. On the fifth card! I looked over at Ram in shock. He'd blown it. He'd overreached. Maurice, sitting in the front row, smacked his forehead.

"We have a new United States memory champion!"

I didn't stand up. I'm not even sure I breached a smile. A minute earlier, all I had wanted was to win. But now my first emotion was not happiness or relief or self-congratulation. It was, I was surprised to discover, simply exhaustion. I felt the sleeplessness of the previous night wash over me, and kept my head buried in my hands for a moment. People watching at home probably thought I was overcome with emotion. In fact, I was still stuck inside my memory palace, floating through a world of impossible images that seemed for a brief moment more real than the stage I was sitting on. I looked up and saw the kitschy, two-tiered trophy twinkling at the edge of the stage. Ram reached over to shake my hand and whispered in my ear, "The fifth card. What was it?"

I dropped my hands, turned to him, and whispered back: "The five of clubs." Dom DeLuise. Hula-hooping. Of course.

EPILOGUE

"Congratulations to Joshua Foer. He's really going to have a story to write about this time, isn't he?" announced the play-by-play man Kenny Rice. "He came here really just to see what it would take and he's going home a champion."

"Well, not bad for a rookie, Joshua," said Ron Kruk, the HDNet reporter who had ascended the stage with a microphone in hand for a postgame interview. "You came in and covered this event a couple times. How key was that experience in your becoming so successful and winning the USA Memory Championship today?"

"I think it was important but I think the practice I put in leading up to today was probably more important," I said.

"Well, it paid off for you today, definitely. You're on your way to the world championships."

That absurd thought hadn't even occurred to me.

"You've been there and covered that as a journalist. How is that going to help you?"

I laughed. "I don't have any chance in the world championships, to be honest. Those people can memorize a deck of cards in, like, thirty seconds. They're extraterrestrials, basically."

"I'm sure you'll do the United States proud. We're all counting on you. You know, if you win the Super Bowl, you say, 'I'm going to Disneyland.' If you win the USA Memory Championship, you say . . ."

He shoved the microphone in my face. I was supposed to answer that I was going to Kuala Lumpur, I guess. Or maybe I was supposed to say Disneyland. I was confused. And very, very tired. And the cameras were rolling. "Um. I don't know," I said. I was at a loss. "I think I'm going home."

As soon as I got off the podium, I rang Ed from the nearest pay phone. It was mid-morning in Australia, and he was standing in the outfield of a cricket pitch, engaging, he said, in a bit of "experimental philosophy."

"Ed, it's Josh—"

"Did you win?" The words rushed out of his mouth as if he'd been waiting all morning for my call.

"I won."

He let out a roar. "What a spectacular coup! Well done, man, well done! You know what this means, right? You are now the undisputed owner of the brains of America!"

The next morning, out of curiosity, I went to the memory circuit's online bulletin board to see if the full scores from the competition had been posted yet, and what, if anything, the Europeans had to say about a novice having bested the American field. Ben had already written up a fourteen-page report on the championship. The last section included a few words on the new champion: "I was impressed with his performance, considering how short a time he's been training, and I think he might just be the person who takes American memory competitions to new heights," Ben wrote. "He's learned his techniques from Europeans, he's been to the competitions over here, and he's not restricted like the others by the low standards necessary to make it big in America. He's got a genuine passion for the sport, and I think he could go on to be not just a grand master, but maybe the first American to get into the top echelon of memory competitors. And when he does, no doubt his countrymen will up their game to keep up with him. It only takes one person to inspire others. So I think the future looks bright for memory in America!"

The USA memory champion turns out to be a minor (OK, very minor) celebrity. All of a sudden, Ellen DeGeneres wanted to talk to me, and *Good Morning America* and the *Today* show were calling to ask if I'd memorize a deck of cards on the air. ESPN wanted to know if I'd learn the NCAA tournament brackets for one of their morning shows. Everyone wanted to see the monkey perform his tricks.

The biggest shock of my newfound stardom (or loserdom, depending on your perspective, I suppose) was that I was now the official representative of all 300 million citizens of the United States of America to the World Memory Championship. This was not a position I had ever expected to be in. At no point during my training did it ever occur to me that I might someday go head-to-head with the likes of Ed Cooke, Ben Pridmore, and Gunther Karsten, the superstars I had initially set out to write about. In all my hours of training, I hardly ever thought to compare my practice scores to theirs. I was a beer league softball right fielder; they were the New York Yankees.

When I showed up in London at the end of August (the championship was moved at the last minute from Malaysia), I brought along my earmuffs, which I'd painted with Captain America stars and stripes;

fourteen decks of playing cards I would try to memorize in the hour cards event; and a Team U.S.A. T-shirt. My principle ambition was simply not to embarrass myself or my country. I also set myself two secondary goals: to finish in the top ten of the thirty-seven-person field and to earn the title of grand master of memory.

As it turned out, both goals were beyond my reach. As the official representative of the greatest superpower on earth, I'm afraid to say I gave the world an entirely mediocre impression of America's collective memory. Though I learned a respectable nine and a half decks of cards in an hour (half a deck short of the grand master standard), my score in hour numbers was a humiliating 380 digits (620 short of grand master). I did manage a third place showing in names and faces, an accomplishment I chalked up to the fact that the packet of names we'd been given to memorize was a veritable United Nations of ethnic monikers. Since I came from the most multicultural country in the world, few of them were unfamiliar to me. Overall, I finished in thirteenth place out of the thirty-seven competitors, behind just about every German, Austrian, and Brit—but, I'm pleased to say, ahead of the French guy, and the entire Chinese team.

On the last afternoon of the championship, Ed took me aside and told me that in recognition of my "fine memory and upstanding character" I would, that night, be offered election into the KL7, provided I could pass the secret society's hallowed initiation ritual.

This gesture, more even than my American championship trophy, signaled true achievement in the world of the memory circuit. I knew that the three-time world champion Andi Bell had never been offered membership in the KL7. Neither had the majority of the world's three dozen grand masters of memory. The only other inductee that year was to be Joachim Thaler, an affable seventeen-year-old Austrian, and he was only invited into the club after placing third in two consecutive world championships. The KL7's membership offer brought my journey full circle in a way I never could have anticipated when I had first set out as an outsider hoping to chronicle the bizarre culture of competitive memorizers. Now I would truly, officially become one of them.

Later that evening, after the young German law student Clemens Mayer wrapped up the world title, and after the awards ceremony at which a bronze medal was placed around my neck for my third-place finish in the names-and-faces event, the entire memory circuit gathered for a celebratory dinner at Simpson's-in-the-Strand, the grand old restaurant where the greatest chess players of nineteenth-century London used to gather, and where one of the most legendary chess matches of all time, the "Immortal Game" of 1851, was played by Adolf Anderssen and Lionel Kieseritzky. Several members of the KL7 ducked out before dessert and congregated in the lobby of charter member Gunther Karsten's hotel down the street.

Ed, who had traveled across town wearing two silver medals around his neck (for his sixteen decks in the hour cards event and 133 consecutive digits in spoken numbers), sat down in a leather chair next to me, under a large carved stone fireplace. "Let me lay this out for you," he said. "In order to join our ranks, you will need to accomplish the following three tasks within five minutes: You will have to drink two beers, memorize forty-nine digits, and kiss three women. Do you understand the task before you?"

"I do."

Gunther paced back and forth behind me in a skin-tight undershirt.

"This is eminently doable, Josh," Ed said, removing his watch from his wrist. "We're going to give you one minute of preparation time to decide if you want to down the beers before you memorize or while you memorize. But as a cautionary tale, let me inform you that someone once tried to memorize the forty-nine digits, and then drank the two pints immediately before recall, and he is not yet a member of the KL7." He looked down at his watch. "Either way, the clock starts ticking when I say go."

One of the mental athletes, who was not in the KL7 but who had tagged along to the induction ceremony, scribbled out forty-nine digits on the back of a business card. Ed screamed, "Go," and I cupped my hands around my ears as makeshift muffs and started memorizing: 7 . . . 9 . . . 3 . . . 8 . . . 2 . . . 6 . . . I took a big gulp of beer between every sixth digit. Just as I finished etching an image of the final two digits, Ed called out, "Time!" and stripped the numbers out of my hand.

I lifted my head out of my hands, and started smoothly listing off digits. But when I got to the last locus of my memory palace, I found my image of the final two digits had evaporated. I ran through every possible digit combination from 00 to 99, but none of them fit. I opened my eyes and begged for a hint. There was silence.

"I didn't make it, did I?"

"No, I'm sorry, forty-seven digits will not suffice," Ed solemnly pronounced to the assembled members of the club. He turned back to address me. "I'm really sorry."

"Don't worry, I didn't make it my first time either," said Gunther, patting me on the shoulder.

"Does this mean I'm not in the KL7?"

Ed tightened his lips and shook his head. His response was surprisingly stern. "No, Josh. You're not."

"Please, Ed, isn't there something you can do?" I pleaded.

"I'm afraid friendship is getting in the way of KL7 business. If you want to become a member of our club, you're going to have to start over." He beckoned for the waitress. "Believe me, it's much more impressive to get in to the KL7 the farther along into the evening you go."

A new table of forty-nine digits was drawn up, and two more pints were poured. This time, miraculously, my images were as clear as any

I'd created all weekend—and twice as obscene. And unlike my first go-round, I even had enough spare time to take one extra walk through my palace. When Ed called time, I closed my eyes and read off the forty-nine digits as confidently as if I'd been practicing them all day.

Ed stood up and gave me a high five and a hug. But Gunther, who was by now, like me, quite drunk, was not appeased. He insisted on one last hurdle before I could be officially inducted into the KL7. "You must still kiss three times the knee of a strange woman," he said.

"One knee? Three times? Now you're just making the rules up as we go," I protested.

"This is how it is," he said.

He took me by the arm and pulled me into an adjoining room of the bar, where he tried to explain the situation to a pair of middle-aged Irish women who were quietly enjoying glasses of wine. I seem to remember telling one of them not to worry, that there was nothing at all weird about the situation: We were memory champions, and this was actually quite an honor for her knee. I also seem to recall that line of logic not working, but Gunther coming up with something even more persuasive. Somehow I ended up on one knee giving three pecks to some poor woman's bare kneecap, after which Gunther hoisted my arm into the air and declared that I had met every challenge, passed every test, and deserved admission to the world's most esteemed organization of mental athletes. "Welcome to our great club KL7!" he shouted.

My memories of the rest of that evening are splotchy. I remember sitting with Tony Buzan on a couch and repeatedly telling him that he was "the Man," while ostentatiously winking at Ed over his shoulder. I remember Ben joking that the waitress must have thought we were all a bunch of weirdos. I remember Ed telling me that "our friendship is epic."

Looking back at my reporter's notebook from that night, the gradual diminishment of my mental state is obvious. Over the course of the evening my handwriting starts to scrawl across the page. It is barely legible today, though one page is clear enough: "Holy Crap! I'm in the KL7! And I think I'm in the Women's Bathroom!"

On the facing page of my notebook, the handwriting all of a sudden becomes clean again, and transitions into the third person. I had become too inebriated to write, and was having too much fun in any case. I handed off my notebook to the nearest sober person I could find, and told her to try to be objective. There was no point pretending I was still a journalist.

Having spent the better part of a year trying to improve my memory, I returned to Florida State University to spend another day and a half being retested by Anders Ericsson and his grad students Tres and Katy in the same cramped office where almost a year earlier I had undergone a top-to-bottom examination of my memory. With Tres once again looking over my shoulder, and a head-mounted microphone once again dangling before my mouth, I retook the same battery of tests, as well as a handful of new ones.

So had I improved my memory? By every objective measure, I had improved something. My digit span, the gold standard by which working memory is measured, had doubled from nine to eighteen. Compared with my tests almost a year earlier, I could recall more lines of poetry, more people's names, more pieces of random information thrown my way. And yet a few nights after the world championship, I went out to dinner with a couple of friends, took the subway home, and only remembered as I was walking in the door to my parents' house that I'd driven a car to dinner. I hadn't just forgotten where I parked it. I'd forgotten I had it.

That was the paradox: For all of the memory stunts I could now perform, I was still stuck with the same old shoddy memory that misplaced car keys and cars. Even while I had greatly expanded my powers of recall for the kinds of structured information that could be crammed into a memory palace, most of the things I wanted to remember in my everyday life were not facts or figures or poems or playing cards or binary digits. Yes, I could now memorize speeches, and the names of dozens of people at a cocktail party, and that was surely useful. And you could give me a family tree of English monarchs, or the terms of the American secretaries of the interior, or the dates of every major battle in World War II, and I could learn that information relatively fast, and even hold on to it for a while. These skills would have been a godsend in high school. But life, for better or worse, only occasionally resembles high school.

While my digit span may have doubled, could it really be said that my working memory was twice as good as it had been when I started my training? I wish I could say it was. But the truth is, it wasn't. When asked to recall the order of, say, a series of random inkblots or a series of color swatches or the clearance of the doorway to my parents' cellar, I was no better than average. My working memory was still limited by the same magical number seven that constrains everyone else. Any kind of information that couldn't be neatly converted into an image and dropped into a memory palace was just as hard for me to retain as it had always been. I'd upgraded my memory's software, but my hardware seemed to have remained fundamentally unchanged.

And yet clearly I had changed. Or at least how I thought about myself had changed. The most important lesson I took away from my year on the competitive memory circuit was not the secret to learning poetry by heart, but rather something far more global and, in a way, far more likely to be of service in my life. My experience had validated the old saw that practice makes perfect. But only if it's the right kind of concentrated, self-conscious, deliberate practice. I'd learned firsthand that with focus, motivation, and, above all, time, the mind can be trained to do extraordinary things. This was a tremendously empowering discovery. It made me ask myself: What else was I capable of doing, if only I used the right approach?

Once our testing had wrapped up, I asked Ericsson whether he thought anyone who'd put in the same amount of time as I did could have improved his memory to the degree that I had.

"I think that with only one data point, we don't know," he told me. "But it's rare for someone to make the kind of commitment you made, and I think your willingness to take on the challenge may make you different. You're clearly not a random person, but on the other hand, I'm not sure there's anything in how you improved that is completely outside the range of what a motivated college student could do."

When I started on this journey, standing with my journalist's notebook in the back of the Con Edison auditorium more than a year earlier, I didn't know where it would lead, *how* thoroughly it would take over my life, or how it would eventually alter me. But after having learned how to memorize poetry and numbers, cards and biographies, I'm convinced that remembering more is only the most obvious benefit of the many months I spent training my memory. What I had really trained my brain to do, as much as to memorize, was to be more mindful, and to pay attention to the world around me. Remembering can only happen if you decide to take notice.

The problem that bedeviled the synesthete S and the fictional Funes was an inability to distinguish between those details that were worth paying attention to and those that weren't. Their compulsive remembering was clearly pathological, but I can't help but imagine that their experience of the world was also, perversely, richer. Nobody would want to have their attention captured by every triviality, but there is something to be said for the value of not merely passing through the world, but also making some effort to capture it—if only because in trying to capture it, one gets in the habit of noticing, and appreciating.

I confess that I never got good enough at filling memory palaces on the fly to feel comfortable throwing out my Dictaphone and notebook. And as someone whose job requires knowing a little bit about a lot, my reading habits are necessarily too extensive to be able to practice more than the occasional intensive reading and memorizing that Ed preaches. Though I committed quite a few poems to heart using memory techniques, I still haven't tackled a work of literature longer than "The Love Song of J. Alfred Prufrock." Even once I'd reached the point where I could squirrel away more than thirty digits a minute in memory palaces, I still only sporadically used the techniques to memorize the phone numbers of people I actually wanted to call. I found it was just too simple to punch them into my cell phone. Occasionally, I'd memorize shopping lists, directions, or to-do lists, but only in the rare circumstances when there wasn't a pen available to jot them down. It's not that the techniques didn't work. I am walking proof that they do. It's that it is so hard to find occasion to use them in the real world in which paper, computers, cell phones, and Post-its can handle the task of remembering for me.

So why bother investing in one's memory in an age of externalized memories? The best answer I can give is the one that I received unwittingly from EP, whose memory had been so completely lost that he could not place himself in time or space, or relative to other people. That is: How we perceive the world and how we act in it are products of how and what we remember. We're all just a bundle of habits shaped by our memories. And to the extent that we control our lives, we do so by gradually altering those habits, which is to say the networks of our memory. No lasting joke, invention, insight, or work of art was ever produced by an external memory. Not yet, at least. Our ability to find humor in the world, to make connections between previously unconnected notions, to create new ideas, to share in a common culture: All these essentially human acts depend on memory. Now more than ever, as the role of memory in our culture erodes at a faster pace than ever before, we need to cultivate our ability to remember. Our memories make us who we are. They are the seat of our values and source of our character. Competing to see who can memorize more pages of poetry might seem beside the point, but it's about taking a stand against forgetfulness, and embracing primal capacities from which too many of us have become estranged. That's what Ed had been trying to impart to me from the beginning: that memory training is not just for the sake of performing party tricks; it's about nurturing something profoundly and essentially human.

Before the KL7 festivities degraded into a debauched free-for-all of blindfolded chess games and drunken recitations of the previous day's poem, Gunther cornered me on a couch and asked if I would continue competing on the memory circuit. I told him that a not small part of me wanted to keep it up. It was, after all, not only strangely thrilling in a way I could have never predicted, but also addictive. That night, I could envision something I'd never before contemplated: the possibility of getting sucked in even deeper. After all, I had a USA title and a speed cards record to defend, and I was sure I could break the minute barrier in cards if I only put in a bit more time. Not to mention historic dates: I could do so much better in historic dates! And there was the grand master standard I'd just missed. "'Grand Master of Memory' would look awfully nice on a business card," I joked to Gunther (it actually is on his business card). I could have filled a memory palace with the scenes I was imagining: the millennium system I'd develop, the horse blinders I'd buy, the hours of practice I'd invest, the jet-setting to national championships around the world. But even then, at the very moment I was being offered admission to the memory circuit's sanctum sanctorum, I was sober enough to recognize that it was time for me to hang up my cleats. My experiment was over. The results were in. I told Gunther that I would miss it, but I didn't see myself coming back next year.

"It's too bad," he said, "but I understand it. It would mean a lot more practice, and that's time which you very likely can invest in a much better way." He was right, I thought. I wondered why he'd never managed to have that realization about himself.

Ed got up off the couch and raised a toast to me, his star pupil. "Let's go get a bagel," he said, and we walked out the door. I have no memory of the rest of the night. I woke up the next afternoon with a large red circle on my cheek—the imprint of my names-and-faces bronze medal. I'd forgotten to take it off.

ACKNOWLEDGMENTS

This book took a while. I'm grateful to everyone who supported me in its creation as readers of drafts, sources of expertise, proofreaders, and friends. There are more of you than I could possibly name. I am especially grateful to all of the mental athletes who spent so much time with me, generously sharing their knowledge and their lives.

This book benefited from two editors. Vanessa Mobley guided it through its initial stages. Eamon Dolan expertly saw it through to completion. I am grateful to Ann Godoff for her faith in me and to everyone at Penguin Press for their work on this book's behalf. My literary agent, Elyse Cheney, is the best partner anyone could ask for. Lindsay Crouse was an extraordinary checker of hard-to-pin-down facts. Brendan Vaughan helped make my writing much sharper.

In the interests of explanatory expediency, I have moved some details, conversations, and scenes around chronologically, but these changes don't materially affect the truth of this book. To the extent that memory records and other time-sensitive facts are not always up-to-date, that is because I have tried to tell this story from the perspective I had when originally experiencing it. In the three years it took me to write this book, much changed in the world. My girlfriend became my wife. The thirty-second barrier in speed cards fell, and fell again. The poem event was finally nixed from international competition. And sadly, EP and Kim Peek passed away. I feel profoundly lucky for the time I was able to spend with them.

NOTES

1: THE SMARTEST MAN IS HARD TO FIND

12 **$265 million industry in 2008:** *Sharp Brains Report* (2009).

2: THE MAN WHO REMEMBERED TOO MUCH

27 **80 percent of what they'd seen:** Lionel Standing (1973), "Learning 10,000 Pictures," *Quarterly Journal of Experimental Psychology* 25, 207–22.

27 **2,500 images:** Timothy F. Brady, Talia Konkle, et al. (2008), "Visual Long-Term Memory Has a Massive Storage Capacity for Object Details," *Proceedings of the National Academy of Sciences* 105, no. 38, 14325–29.

28 **"details could eventually be recovered":** Elizabeth Loftus and Geoffrey Loftus (1980), "On the Permanence of Stored Information in the Human Brain," *American Psychologist* 35, no. 5, 409–20.

28 **Wagenaar came to believe the same thing:** Willem A. Wagenaar (1986), "My Memory: A Study of Autobiographical Memory over Six Years," *Cognitive Psychology* 18, 225–52.

30 **only one case of photographic memory has ever been described in the scientific literature:** Photographic memory is often confused with another bizarre—but real—perceptual phenomenon called eidetic memory, which occurs in 2 to 15 percent of children, and very rarely in adults. An eidetic image is essentially a vivid afterimage that lingers in the mind's eye for up to a few minutes before fading away. Children with eidetic memory never have anything close to perfect recall, and they typically aren't able to visualize anything as detailed as a body of text. In these individuals, visual imagery simply fades more slowly.

30 **a paper in *Nature*:** C. F. Stromeyer and J. Psotka (1970), "The Detailed Texture of Eidetic Images," *Nature* 225, 346–49.

30 **none of them could pull off Elizabeth's nifty trick:** J. O. Merritt (1979), "None in a Million: Results of Mass Screening for Eidetic Ability," *Behavioral and Brain Sciences* 2, 612.

31 **"other people having photographic memory":** If anyone alive today has a claim to photographic memory, it's a British savant named Stephen Wiltshire, who has been called the "human camera" for his ability to create sketches of a scene after looking at it for just a few seconds. But even he doesn't have a truly photographic memory, I learned. His mind doesn't work like a Xerox machine. He takes liberties. And curiously, his cameralike abilities extend only to drawing certain kinds of objects and scenes, namely architecture and cars. He can't, say, look at a page of the dictionary and then instantly recall what was on it. In every case except Elizabeth's where someone has claimed to have a photographic memory, there has always been another explanation.

31 **"none of them ever attained any prominence in the scholarly world":** George M. Stratton (1917), "The Mnemonic Feat of the 'Shass Pollak,'" *Psychological Review* 24, 244–47.

33 **a pattern of connections between those neurons:** Recently, a paper in the journal *Brain and Mind* attempted to estimate the capacity of the human brain using a model that treats a memory as something stored not in individual neurons but rather in the connections between neurons. The authors estimated that the human brain can store 10^{8432} bits of information. By contrast, it's said that there are somewhere on the order of 10^{78} atoms in the observable universe.

38 **physically altered the gross structure of their brains:** E. A. Maguire et al. (2000), "Navigation-Related Structural Change in the Hippocampi of Taxi Drivers," *PNAS* 97, 84398–403.

39 **not a single significant structural difference turned up:** E. A. Maguire, et al (2003), "Routes to Remembering: The Brains Behind Superior Memory," *Nature Neuroscience* 6 no.1, 90–95.

40 **wouldn't seem to make any sense:** If the mental athletes were also using navigational skills, why didn't they have enlarged hippocampuses, like the taxi drivers? The likely answer is that mental athletes simply don't use their navigational abilities nearly as much as taxi drivers.

44 **"Baker/baker paradox":** G. Cohen (1990), "Why Is It Difficult to Put Names to Faces?" *British Journal of Psychology* 81, 287–97.

3: THE EXPERT EXPERT

49 **all the hard work of putting food on our tables:** I'm speaking here about egg-laying chickens, which are distinct from broiler chickens bred to produce meat.

52 **"Exceptional Memorizers: Made, Not Born":** K. Anders Ericsson (2003), "Exceptional Memorizers: Made, Not Born," *Trends in Cognitive Sciences* 7, no.6, 233–35.

53 **volleyball defenders:** Much of this research is captured in *The Cambridge Handbook of Expertise and Expert Performance*, edited by K. Anders Ericsson, Neil Charness, Paul J. Feltovich, and Robert R. Hoffman.

65 **several opponents at once, entirely in their heads:** During the first half of the twentieth century, playing simultaneous games of blindfolded chess against multiple opponents became a fetishized skill in the chess world. In 1947, an Argentinian grand master named Miguel Najdorf set a record by playing forty-five simultaneous games in his mind. It took him twenty-three and a half hours, and he finished with a record of thirty-nine wins, four losses, and two draws, and then was unable to fall asleep for three straight days and nights afterward. (According to chess lore, simultaneous blindfolded chess was once banned in Russia due to the mental health risks.)

4: THE MOST FORGETFUL MAN IN THE WORLD

69 **lab technician called EP:** L. Steffanaci et al. (2000), "Profound Amnesia After Damage to the Medial Temporal Lobe: A Neuroanatomical and Neuropsychological Profile of Patient E. P.," *Journal of Neuroscience* 20, no. 18, 7024–36.

5: THE MEMORY PALACE

94 **textbook called the *Rhetorica ad Herennium*:** So named after Gaius Herennius, the book's patron.

94 **"This book is our bible":** The little red Loeb Classical Library English/Latin edition of the book has the Roman statesman and philosopher Cicero's name printed on its spine—albeit inside a pair of brackets. Until at least the fifteenth century, people believed the short treatise had been written by the great Roman orator himself, but modern scholars have long been doubtful. It made sense that Cicero would have been the book's author, since he was not only a famous master of memory techniques—he delivered his legendary speeches before the Roman senate from memory—but also (definitively) the author of another work called *De Oratore*, which is where the story of Simonides and the banquet hall first appeared. That the story of Simonides, a fifth-century-B.C. Greek, would have its first written record in a book written four centuries afterward by a Roman reflects the fact that no memory treatises have survived from ancient Greece—though some must certainly have been written. Since Cicero's recounting of the incident was written so much later than Simonides supposedly remembered the locations of the mangled bodies, nobody can say just how much of the story is myth. I'm willing to wager that quite a lot of it is mythical, but a marble tablet dating to 264 B.C.—two centuries before Cicero, but still two centuries after the fact—and unearthed in the seventeenth century describes Simonides as "the inventor of the system of memory aids." Still, it's hard to believe that a technique like the art of memory was invented by one person at one moment in time, in so perfectly poetic a manner. For all we know, Simonides was merely the art of memory's codifier, or maybe just a particularly adept practitioner who got tagged as its inventor. In any case, Simonides was a real person, and a real poet—the first apparently to charge for his poems and also the first to have called poetry "vocal painting" and painting "silent poetry." This is a particularly noteworthy turn of phrase for Simonides to have coined because the art of memory that he is credited with inventing is all about turning words into paintings in the mind.

100 **less a test of memory than of creativity:** The key thing is to compress as much information as possible into any single well-formed image. The *Ad Herennium* gives the example of a lawyer who needs to remember the basic facts of a case: "The prosecutor has said that the defendant killed a man by poison, has charged that the motive for the crime was an inheritance,

and declared that there are many witnesses and accessories to this act." To remember all this, "we shall picture the man in question as lying ill in bed, if we know his person. If we do not know him, we shall yet take some one to be our invalid, but not a man of the lowest class, so that he may come to mind at once. And we shall place the defendant at the bedside, holding in his right hand a cup, and in his left, tablets, and on the fourth finger a ram's testicles." The bizarre image would certainly be tough to forget, but it takes some decoding to figure out exactly what it is you're supposed to be remembering. The cup is a mnemonic to remind us of the poison, the tablets are a reminder of the will, and the ram's testicles are a double entendre, reminding us of the witnesses with a verbal pun on *testes* (testimony) and—since Roman purses were often made out of the scrotum of a ram—of the possibility of bribing them. Seriously.

100 **"memory is marvelously excited by images of women":** Rossi, *Logic and the Art of Memory*, p. 22.

6: HOW TO MEMORIZE A POEM

110 **" judgment, citizenship, and piety":** Carruthers, *The Book of Memory*, p. 11.

110 **"worth a thousand in the stacks":** Draaisma, *Metaphors of Memory*, p. 38.

110 **the principle language in which he wrote:** Carruthers, *The Book of Memory,* p. 88.

125 **"core of his educational equipment":** Havelock, *Preface to Plato*, p. 27.

125 **Professional memorizers:** My favorite story about professional memorizers is told by Seneca the Younger about a wealthy Roman aristocrat named Calvisius Sabinus, who gave up on trying to learn the great works by heart and instead hired a coterie of slaves to do it for him.

> I never saw a man whose good fortune was a greater offence against propriety. His memory was so faulty that he would sometimes forget the name of Ulysses, or Achilles, or Priam . . . But nonetheless did he desire to appear learned. So he devised this shortcut to learning: he paid fabulous prices for slaves—one to know Homer by heart and another to know Hesiod; he also delegated a special slave to each of the nine lyric poets. You need not wonder that he paid high prices for these slaves . . . After collecting this retinue, he began to make life miserable for his guests; he would keep these fellows at the foot of his couch, and ask them from time to time for verses which he might repeat, and then frequently break down in the middle of

a word . . . Sabinus held to the opinion that what any member of his household knew, he himself also knew.

125 **memorizing the Vedas with perfect fidelity:** The Rigveda, the oldest of the Vedic texts, is over ten thousand verses long.

125 **attached to poets as official memorizers:** After the introduction of Islam, Arabic mnemonists became known as *huffaz*, or "holders" of the Koran and Hadith.

125 **memorized the oral law on behalf of the Jewish community:** For more on Jewish mnemonists, see Gandz, "The Robeh, or the Official Memorizer of the Palestinian Schools."

126 **gathering armies, heroic shields, challenges between rivals:** Ong, *Orality and Literacy*, p. 23, and Lord, *The Singer of Tales*, pp. 68–98.

126 **that was about as far as his inquiry into the matter went:** As it turns out, this radical argument was actually not new at all. In fact, it seems long ago to have been a widely accepted notion that was somehow forgotten. The first century A.D. Jewish historian Josephus wrote, "They say that even Homer did not leave his poetry in writing, but that it was transmitted by memory." And according to a tradition repeated by Cicero, the first official redaction of Homer was ordered by the Athenian tyrant Peisistratus in the sixth century B.C. As people's connections to oral culture grew more distant over the centuries, the idea of literature without writing became a harder and harder notion to digest, and eventually just came to seem implausible.

127 **"composed wholly without the aid of writing":** For more, see Ong, *Orality and Literacy*, which is a major source for this chapter.

129 **"Word for word, and line for line":** As reported by Parry's student Albert Lord in *The Singer of Tales*, p. 27.

130 **before trying to see it as a series of images:** Carruthers argues in a revised second edition of *The Book of Memory* that the *memoria verborum* has long been misunderstood by modern psychologists and scholars. It was not, in fact, an alternative to rote, verbatim memorization, she contends, and was never meant to be used for memorizing long stretches of text. Rather, she suggests, it was for recalling single words and phrases—perhaps as long as a line of verse—that one had trouble remembering accurately.

131 **the quandary of how to see the unseeable:** According to Pliny, it was Simonides who invented the art of memory but Metrodorus who perfected it. Cicero called the man "almost divine."

132 ***balistarius*:** Alternatively, Bradwardine's system allowed that you could reverse a syllable simply by imagining an image upside-down, so "ba-" could also just be an abbot hanging from the ceiling.

132 **an abbot getting shot by a crossbow:** Or an abbot having a conversation with another abbot who was hanging from the ceiling.

132 **"mangles or caresses St. Dominic":** Carruthers, *The Book of Memory*, pp. 136–37.

132 **depraved carnal affections:** Yates, *The Art of Memory*, p. 277.

7: THE END OF REMEMBERING

139 **that we have any knowledge of it today:** Manguel, *A History of Reading*, p. 60.

139 **a time when writing was ascendant in Greece:** In Socrates' day, about 10 percent of the Greek world was literate.

140 **"in material books to help the memory":** Carruthers, *The Book of Memory*, p. 8.

140 **some stretching up to sixty feet:** Fischer, *A History of Writing*, p. 128.

140 **papyrus reeds imported from the Nile Delta:** Papyrus, the literal bulrushes of the biblical "ark of bulrushes" that carried the baby Moses, was also called *byblos*, after the Phoenician port of Byblos where it was exported—hence the "Bible." In the second century B.C., the Hellenistic ruler of Egypt, Ptolemy V Epiphanes, cut off papyrus exports in order to curtail the growth of a rival library at Pergamum in Asia Minor (the word "parchment"—derived from *charta pergamena*—is a tribute to Pergamum, where the material was used extensively). From then on, it became more common for books to be penned on stretched parchment or vellum (a final piece of ancient book etymology: vellum, which was often made from calfskin, shares the same root with "veal"), both of which lasted longer and were more transportable than papyrus.

140 **how long to pause between sentences:** He created the high point, ˙, corresponding to the modern period, the middle point, ; , corresponding to the modern semicolon, and the low point, ., corresponding to the modern comma. The middle point vanished in the Middle Ages. The question mark didn't appear until the publication of Sir Philip Sydney's *Arcadia* in 1587, and the exclamation mark was first used in the Catechism of Edward VI in 1553.

141 **GREECE:** Small, *Wax Tablets of the Mind*, p. 53. I've borrowed her idea of printing English in this manner to show how hard it is to read.

141 **a phrase often repeated in medieval texts:** For more on reading *scriptio continua*, see Manguel, *A History of Reading*, p. 47.

142 **extremely difficult to sight-read:** Indeed, much published modern Hebrew, like the kind you'd find in a newspaper in Tel Aviv, is written without vowels. Words generally have to be recognized as units, rather than sounded out as they are in English. This slows Hebrew readers down. Native Hebrew speakers who also read English can typically read English translations faster than their own native language, even though it takes about 40 percent more words to say the same thing in English as in Hebrew.

143 **"The stuff he knows made him lick her":** Sounds that can be sliced up in different ways to yield different semantic meanings are known as oronyms. The "stuffy nose" comes from Pinker, *The Language Instinct*, p. 160.

143 **a giant and very curious step backward:** Small, *Wax Tablets of the Mind*, p. 114.

143 ***ánagignósko*:** Carruthers, *The Book of Memory*, p. 30.

143 **ten billion volumes:** Man, *Gutenberg: How One Man Remade the World*, p. 4.

143 **would have been considered particularly well stocked:** In 1290, the library at the Sorbonne, among the biggest in the world, held exactly 1,017 books—fewer titles than many readers of this book will devour in a lifetime.

144 **hadn't even been invented yet:** For more on the history of the display of books, see Petroski, *The Book on the Bookshelf*, pp. 40–42.

144 **still weighed more than ten pounds:** Illich, *In the Vineyard of the Text*, p. 112.

144 **around the same time that chapter divisions were introduced:** *The Comprehensive Concordance to the Holy Scriptures* (1894), pp. 8–9.

144 **reading the text all the way through:** Draaisma, *Metaphors of Memory*, p. 34.

145 **"pre- and post-index Middle Ages":** Illich, *In the Vineyard of the Text*, p. 103.

145 **labyrinthine world of external memory:** A point made by Draaisma in *Metaphors of Memory*.

146 **"living concordance":** In the words of Carruthers, *The Craft of Thought*, p. 31.

146 **how to memorize playing cards:** Corsi, *The Enchanted Loom*, p. 21.

147 **"the letter A":** Translation quoted from Carruthers, *The Book of Memory*, p. 114.

147 **"intensive" to "extensive" reading:** Darnton attributes this idea to Rolf Engelsing, who cites the transformation as happening as late as the eighteenth century. *The Kiss of Lamourette*, p. 165.

149 **one of the most famous men in all of Europe:** Yates's assessment in *The Art of Memory*, p. 129.

149 **round, seven-tiered edifice:** Yates tried to reconstruct the blueprints for the theater in *The Art of Memory*.

150 **"and all the things that are in the entire world":** Rossi, *Logic and the Art of Memory*, p. 74.

150 **hundreds—perhaps thousands—of cards were drafted:** Corsi, *The Enchanted Loom*, p. 23.

150 **over the course of a week:** Much of this information comes from Douglas Radcliff-Ulmstead (1972), "Giulio Camillo's Emblems of Memory," *Yale French Studies* 47, 47–56.

151 **the apotheosis of an entire era's ideas about memory:** More recently, virtual reality gurus have come to see Camillo's memory theater as the historical forerunner of their entire field—and have traced its influence all the way to the Internet (the ultimate universal memory palace) and the Apple and Windows operating systems, whose spatially arranged folders and icons are just a modern reworking of Camillo's mnemonic principles. See Peter Matussek (2001), "The Renaissance of the Theater of Memory," *Janus* 8 *Paragrana* 10, 66–70.

152 **"riding a sea monster":** These translations are from Rowland, *Giordano Bruno*, pp. 123–24.

152 **"a parrot on his head":** Eco, *The Search for the Perfect Language*, p.138.

153 **nine pairs of cranial nerves:** There are now twelve known pairs of cranial nerves.

153 **almost a half million dollars:** Fellows and Larrowe, *Loisette Exposed*, p. 217.

154 **a memory course lasting several weeks:** Walsh and Zlatic (1981), "Mark Twain and the Art of Memory," *American Literature* 53, no. 2, 214–31.

8: THE OK PLATEAU

164 **Johann Winkelmann:** The German philosopher Gottfried Leibniz also wrote about a similar system in the seventeenth century, but it's quite likely that the idea of making numbers more memorable by turning them into words was discovered much earlier. The Greeks had an acrophonic system, wherein the first letter of each numeral could be used to represent the number, so that, for example, P represented the number five, for *penta*. In Hebrew, each letter of the *aleph bet* corresponds to a number, a quirk that Kabbalists have used to seek out hidden numerical meanings in Scripture. Nobody knows whether these systems were ever used to memorize numbers, but it's hard to imagine

that some Mediterranean businessman who had to do mental accounting wouldn't have stumbled onto such an obvious idea.

166 **advance the sport of competitive memory by a quantum leap:** Ed gave me the following example of his Millennium PAO system at work: "The number 115 is Psmith, the stylish character from the P. G. Wodehouse books (the P is silent, by the way, as in 'phthisis' or 'ptarmigan'). His action is that he gives away an umbrella that doesn't belong to him to a delicate young lady he sees stranded in a rainstorm. The number 614 is Bill Clinton, who smokes but does not inhale marijuana, and the number 227 is Kurt Gödel, the obsessive logician, who starved himself to death by accident because he was too busy doing formal logic. Now, I can combine these three numbers to form nine-digit numbers that have anecdotal coherence. For example, 115,614,227 becomes Psmith deigning to puff at—without going so far as to inhale—formal logic. Now this is quite understandable since logic is, after all, an activity unsuited to the true English gent. If you change the ordering of the numbers, you get a different anecdote. The number 614,227,115 becomes Bill Clinton mortally forgetting to eat because he's too busy pinching umbrellas for pretty young girls. This image will interact with my pre-existing knowledge of Clinton's life—seeing as how he has gotten into trouble before with the inappropriate handling of cylindrical objects for young ladies—and the chance activation of this association, and the glimmer of accompanying humor, serves to better the stability of the memory. See, each possible combination has its own dynamic feel and emotion, and very often, interestingly, this will be the first thing in recall to pop into one's head, before the other details slowly shuffle into view. I might also mention that this works as an excellent idea-generator and constitutes sound afternoon entertainment."

171 **lesser skaters work more on jumps they've already mastered:** J. M. Deakin and S. Cobley (2003), "A Search for Deliberate Practice: An Examination of the Practice Environments in Figureskating and Volleyball," in *Expert Performance in Sports: Advances in Research on Sport Expertise* (edited by J. L. Starkes and K. A. Ericsson).

172 **trying to understand the expert's thinking at each step:** K. A. Ericsson, et al. (1993), "The Role of Deliberate Practice in the Acquisition of Expert Performance," *Psychological Review* 100 no. 3, 363–406.

172 **working through old games:** N. Charness, R. Krampe, and U. Mayer (1996), "The Role of Practice and Coaching in Entrepreneurial Skill Domains: An

International Comparison of Life-Span Chess Skill Acquisition," in Ericsson, *The Road to Excellence*, pp. 51–80.

173 **have a tendency to get less and less accurate over the years:** C. A. Beam, E. F. Conant, and E. A. Sickles (2003), "Association of Volume and Volume-Independent Factors with Accuracy in Screening Mammogram Interpretation," *Journal of the National Cancer Institute* 95, 282–90.

174 **now acquired by your average high school junior:** Ericsson, *The Road to Excellence*, p. 31.

9: THE TALENTED TENTH

192 **"no sensibilities, no soul":** Ravitch, *Left Back*, p. 21.

193 **"mental discipline":** Ravitch, *Left Back*, p. 61.

203 **inventory and invention:** Carruthers, *The Craft of Thought*, p. 11.

208 **a group of baseball fanatics:** G. J. Spillich (1979), "Text Processing of Domain-Related Information for Individuals with High and Low Domain Knowledge," *Journal of Verbal Learning and Verbal Behavior* 14, 506–22.

208 **either a witch trial or a piece of correspondence:** Frederick M. Hess, *Still at Risk* pp. 1–2.

10: THE LITTLE RAIN MAN IN ALL OF US

215 **meet up with Daniel:** I e-mailed Daniel and asked if he'd be willing to meet with me. He wrote back, "I normally request a fee for interviews with the media." After I explained to him why that would be impossible, he agreed to see me on the condition that I mention the Web site of his online tutoring company, optimnem.co.uk.

219 **legally changed in 2001:** Daniel is fully open about having changed his name. He told me he didn't like the sound of his old family name, Corney,

221 **more than nine thousand books he has read at about ten seconds a page:** It should be noted that this claim was never investigated in a peer-reviewed journal. I suspect this bit of hyperbole might not have held up to careful scrutiny.

226 **it's a skill that can be learned:** Eventually my investigation of mental mathematics led me to a remarkable book called *The Great Mental Calculators: The Psychology, Methods, and Lives of Calculating Prodigies Past and Present* by

a psychologist named Steven Smith. Smith dismisses the notion that there's anything special about the brains of calculation prodigies, and insists that their abilities derive purely from obsessive interest. He compares calculation to juggling: "Any sufficiently diligent non-handicapped person can learn to juggle, but the skill is actually acquired only by a handful of highly motivated individuals." George Packer Bidder, one of the most renowned human calculators of all time, even went so far as to express "a strong conviction, that mental arithmetic can be taught, as easily, if not even with greater facility, than ordinary arithmetic."

230 **would have been able to do as well:** At UCSD, Ramachandran and his graduate students administered three other tests of Tammet's synesthesia. Using Play-Doh, they asked him to create 3-D models of twenty of his number shapes. When they gave him a surprise retest twenty-four hours later, all of his shapes matched up. Then they hooked up an electrode to his fingers and flashed him the digits of pi—but with a few errant digits thrown in. They measured his galvanic skin response and noticed that it jumped dramatically when he confronted a digit that didn't belong.

The UCSD researchers also administered the Stroop test, another assessment commonly used to verify synesthesia. First they gave Daniel three minutes to memorize a matrix of a hundred numbers. After five minutes, he was able to recall sixty-eight of those numbers, and three days later he still remembered them perfectly. Then they gave him three minutes to memorize a matrix of a hundred numbers in which the size of the numbers on the page corresponded to how Daniel described the numbers in his mind. Nines were printed larger than other numbers and sixes were printed smaller. In this case, he memorized fifty digits, and held onto all of them for three days. Finally, they gave him a test where the numbers were printed in incongruous sizes. Nines were printed small. Sixes were printed large. They wanted to see if it would throw Daniel off his game. Did it ever. Daniel was only able to remember sixteen numbers, and after three days, he could remember exactly zero of them. Ramachandran and his students put together a pre-publication conference poster on Daniel titled "Does Synesthesia Contribute to Mathematical Savant Skills?" in which they refer to him by the pseudonym Arithmos. It includes a caveat: "As in all cases like this we need to consider the fact that Arithmos may be performing almost all of his 'mental feats' via pure memorization."

230 **they didn't find this:** D. Bor, J. Bilington, and S. Baron-Cohen (2007), "Savant memory for digits in a case of synaesthesia and Asperger syndrome is related to hyperactivity in the lateral prefrontal cortex." *Neurocase* 13, 311-319.

BIBLIOGRAPHY

Baddeley, A. D. (2006). *Essentials of human memory*. Hove, East Sussex, UK: Psychology Press.

Barlow, F. (1952). *Mental prodigies: an enquiry into the faculties of arithmetical, chess and musical prodigies, famous memorizers, precocious children and the like, with numerous examples of "lightning" calculations and mental magic*. New York: Philosophical Library.

Baron-Cohen, S., Bor, D., Wheelwright, S., & Ashwin, C. (2007). Savant Memory in a Man with Colour Form-Number Synaesthesia and Asperger Syndrome. *Journal of Consciousness Studies*, 14(9-10), 237-251.

Batchen, G. (2004). *Forget me not: photography & remembrance*. New York: Princeton Architectural Press.

Battles, M. (2003). *Library: an unquiet history*. New York: W.W. Norton.

Beam, C. A., Conant, E. F., & Sickles, E. A. (2003). Association of Volume and Volume-Independent Factors with Accuracy in Screening Mammogram Interpretation. *Journal of the National Cancer Institute,* 95, 282-290.

BIBLIOGRAPHY

Bell, C. G., & Gemmell, J. (2009). *Total recall: how the E-memory revolution will change everything.* New York: Dutton.

Bell, G., & Gemmell, J. (2007, March). A Digital Life. *Scientific American,* 58-65.

Biederman, I., & Shiffrar, M. M. (1987). Sexing Day-Old Chicks: A Case Study and Expert Systems Analysis of a Difficult Perceptual-Learning Task. *Journal of Experimental Psychology,* 13(4), 640-645.

Birkerts, S. (1994). *The Gutenberg elegies: the fate of reading in an electronic age.* Boston: Faber and Faber.

Bolzoni, L. (2001). *The gallery of memory: literary and iconographic models in the age of the printing press.* Toronto: University of Toronto Press.

Bolzoni, L. (2004). *The web of images: vernacular preaching from its origins to Saint Bernardino of Siena.* Aldershot, Hants, England: Ashgate.

Bor, D., Billington, J., & Baron-Cohen, S. (2007). Savant memory for digits in a case of synaesthesia and Asperger syndrome is related to hyperactivity in the lateral prefrontal cortex. *Neurocase,* 13(5-6), 311-319.

Bourtchouladze, R. (2002). *Memories are made of this: how memory works in humans and animals.* New York: Columbia University Press.

Brady, T. F., Konkle, T., Alvarez, G. A., & Oliva, A. (2008). Visual Long-Term Memory Has a Massive Storage Capacity for Oject Details. *PNAS,* 105(38), 14325-14329.

Brown, A. S. (2004). *The déjà vu experience.* New York: Psychology Press.

Bush, V. (1945, July). As We May Think. *The Atlantic.*

Buzan, T. (1991). *Use your perfect memory: dramatic new techniques for improving your memory, based on the latest discoveries about the human brain.* New York: Penguin.

Buzan, T., & Buzan, B. (1994). *The mind map book: how to use radiant thinking to maximize your brain's untapped potential.* New York: Dutton.

Caplan, H. (1954). *Ad C. Herennium: de ratione dicendi (Rhetorica ad Herennium).* Cambridge, Mass: Harvard University Press.

Carruthers, M. (1998). *The craft of thought: meditation, rhetoric, and the making of images, 400-1200.* New York: Cambridge University Press.

Carruthers, M. J. (1990). *The book of memory: a study of memory in medieval culture.* Cambridge, England: Cambridge University Press.

Carruthers, M. J., & Ziolkowski, J. M. (2002). *The medieval craft of memory: an anthology of texts and pictures.* Philadelphia: University of Pennsylvania Press.

Cicero, M. T., May, J. M., & Wisse, J. (2001). *Cicero on the ideal orator.* New York: Oxford University Press.

Clark, A. (2003). *Natural-born cyborgs: minds, technologies, and the future of human intelligence.* Oxford, England: Oxford University Press.

Cohen, G. (1990). Why Is It Difficult to Put Names to Faces? *British Journal of Psychology,* 81, 287-297.

Coleman, J. (1992). *Ancient and medieval memories: studies in the reconstruction of the past.* Cambridge, England: Cambridge University Press.

Cooke, E. (2008). *Remember, remember.* London: Viking.

Corkin, S. (2002). What's New with the Amnesic Patient H.M. *Nature Reviews Neuroscience,* 3, 153-160.

Corsi, P. (1991). *The enchanted loom: chapters in the history of neuroscience.* New York: Oxford University Press.

Cott, J. (2005). *On the sea of memory: a journey from forgetting to remembering.* New York: Random House.

Darnton, R. (1990). First Steps Toward a History of Reading. In *The kiss of Lamourette: reflections in cultural history.* New York: W. W. Norton.

Doidge, N. (2007). *The brain that changes itself: stories of personal triumph from the frontiers of brain science.* New York: Viking.

Doyle, B. (2000, March). The Joy of Sexing. *The Atlantic Monthly,* 28-31.

Draaisma, D. (2000). *Metaphors of memory: a history of ideas about the mind.* Cambridge, England: Cambridge University Press.

Draaisma, D. (2004). *Why life speeds up as you get older: how memory shapes our past.* Cambridge, England: Cambridge University Press.

Dudai, Y. (1997). How Big Is Human Memory, or on Being Just Useful Enough. *Learning & Memory,* 3, 341-365.

Dudai, Y. (2002). *Memory from A to Z: keywords, concepts, and beyond.* Oxford, England.: Oxford University Press.

Dudai, Y., & Carruthers, M. (2005). The Janus Face of Mnemosyne. *Nature,* 434, 567.

Dvorak, A. (1936). *Typewriting behavior: psychology applied to teaching and learning typewriting.* New York: American Book Company.

Eco, U. (1995). *The search for the perfect language.* Oxford, England: Blackwell.

Eichenbaum, H. (2002). *The cognitive neuroscience of memory: an introduction.* Oxford, England: Oxford University Press.

Ericsson, K. (2003). Exceptional Memorizers: Made, Not Born. *Trends in Cognitive Science,* 7(6), 233-235.

Ericsson, K. (2004). Deliberate Practice and the Acquisition and Maintenance of Expert Performance in Medicine and Related Domains. *Academic Medicine,* 79(10), 870-881.

Ericsson, K., & Chase, W. G. (1982). Exceptional Memory. *American Scientist,* 70(Nov-Dec), 607-615.

Ericsson, K., & Kintsch, W. (1995). Long-Term Working Memory. *Psychological Review,* 102(2), 211-245.

Ericsson, K. A. (1996). *The road to excellence: the acquisition of expert performance in the arts and sciences, sports, and games.* Mahwah, N.J.: Lawrence Erlbaum Associates.

Ericsson, K. A. (2006). *The Cambridge handbook of expertise and expert performance.* Cambridge, England: Cambridge University Press.

Ericsson, K., Delaney, P. F., Weaver, G., & Mahadevan, R. (2004). Uncovering the Structure of a Memorist's Superior "Basic" Memory Capacity. *Cognitive Psychology,* 49, 191-237.

Ericsson, K., Krampe, R. T., & Tesch-Romer, C. (1993). The Role of Deliberate Practice in the Acquisition of Expert Performance. *Psychological Review,* 100(3), 363-406.

Farrand, P., Hussein, F., & Hennessy, E. (2002). The Efficacy of the 'Mind Map' Study Technique. *Medical Education,* 36(5), 426-431.

Fellows, G. S., & Larrowe, M. D. (1888). *"Loisette" exposed (Marcus Dwight Larrowe, alias Silas Holmes, alias Alphonse Loisette).* New York: G. S. Fellows.

Fischer, S. R. (2001). *A history of writing.* London: Reaktion.

Gandz, S. (1935). The Robeh or the official memorizer of the Palestinian schools. *Proceedings of the American Academy for Jewish Research*, 7, 5–12.

Havelock, E. A. (1963). *Preface to Plato.* Cambridge, Mass.: Belknap Press, Harvard University Press.

Havelock, E. A. (1986). *The muse learns to write: reflections on orality and literacy from antiquity to the present.* New Haven: Yale University Press.

Hermelin, B. (2001). *Bright splinters of the mind: a personal story of research with autistic savants.* London: J. Kingsley.

Herrmann, D. J. (1992). *Memory improvement: implications for memory theory.* New York: Springer-Verlag.

Hess, F. M. (2008). *Still at risk: what students don't know, even now.* Common Core.

Hilts, P. J. (1996). *Memory's ghost: the nature of memory and the strange tale of Mr. M.* New York: Simon & Schuster.

Horsey, R. (2002). *The art of chicken sexing.* Cogprints.

Howe, M. J., & Smith, J. (1988). Calendar Calculating in 'Idiot Savants': How Do They Do It? *British Journal of Psychology,* 79, 371-386.

Illich, I. (1993). *In the vineyard of the text: a commentary to Hugh's Didascalicon.* Chicago: University of Chicago Press.

Jaeggi, S. M., Buschkuehl, M., Jonides, J., & Perrig, W. J. (2008). Improving Fluid Intelligence with Training on Working Memory. *PNAS,* 105(19), 6829-6833.

Johnson, G. (1992). *In the palaces of memory: how we build the worlds inside our heads.* New York: Vintage Books.

Kandel, E. R. (2006). *In search of memory: the emergence of a new science of mind.* New York: W. W. Norton.

Khalfa, J. (1994). *What is intelligence?* Cambridge, England: Cambridge University Press.

Kliebard, H. M. (2002). *Changing course: American curriculum reform in the 20th century.* New York: Teachers College Press.

Kondo, Y., Suzuki, M., Mugikura, S., Abe, N., Takahashi, S., Iijima, T., & Fujii, T. (2005). Changes in Brain Activation Associated with Use of a Memory Strategy: A Functional MRI Study. *NeuroImage,* 24, 1154-1163.

Kurland, M., & Lupoff, R. A. (1999). *The complete idiot's guide to improving your memory.* New York: Alpha Books.

LeDoux, J. E. (2002). *Synaptic self: how our brains become who we are.* New York: Viking.

Loftus, E. F., & Loftus, G. R. (1980). On the Permanence of Stored Information in the Human Brain. *American Psychologist*, 35(5), 409-420.

Loisette, A., & North, M. J. (1899). *Assimilative memory or how to attend and never forget.* New York: Funk & Wagnalls.

Lorayne, H., & Lucas, J. (1974). *The memory book.* New York: Stein and Day.

Lord, A. B. (1960). *The singer of tales.* Cambridge, Mass.: Harvard University Press.

Luria, A. R. (1987). *The mind of a mnemonist: a little book about a vast memory.* Cambridge, Mass.: Harvard University Press.

Lyndon, D., & Moore, C. W. (1994). *Chambers for a memory palace.* Cambridge, Mass.: MIT Press.

Maguire, E. A., Gadian, D. G., Johnsrude, I. S., Good, C. D., Ashburner, J., Frackowiak, R. S., & Frith, C. D. (2000). Navigation-Related Structural Change in the Hippocampi of Taxi Drivers. *PNAS,* 97, 84398-84403.

Maguire, E. A., Valentine, E. R., Wilding, J. M., & Kapur, N. (2003). Routes to Remembering: The Brains Behind Superior Memory. *Nature Neuroscience,* 6(1), 90-95.

Man, J. (2002). *Gutenberg: how one man remade the world with words.* New York: John Wiley & Sons.

Manguel, A. (1996). *A history of reading.* New York: Viking.

Marcus, G. F. (2008). *Kluge: the haphazard construction of the human mind.* Boston: Houghton Mifflin.

Martin, R. D. (1994). *The specialist chick sexer.* Melbourne, Australia.: Bernal Publishing.

Masters of a dying art get together to sex. (2001, February 12). *Wall Street Journal.*

Matussek, P. (2001). The Renaissance of the Theater of Memory. *Janus Paragrana* 8, 66-70.

McGaugh, J. L. (2003). *Memory and emotion: the making of lasting memories.* New York: Columbia University Press.

Merritt, J. O. (1979). None in a Million: Results of Mass Screening for Eidetic Ability. *Behavioral and Brain Sciences,* 2, 612.

Miller, G. A. (1956). The Magical Number Seven, Plus or Minus Two: Some Limits on our Capacity for Processing Information. *Psychological Review,* 63, 81-97.

Mithen, S. J. (1996). *The prehistory of the mind: a search for the origins of art, religion, and science.* London: Thames and Hudson.

Neisser, U., & Hyman, I. E. (2000). *Memory observed: remembering in natural contexts.* New York: Worth.

Noice, H. (1992). Elaborative Memory Strategies of Professional Actors. *Applied Cognitive Psychology,* 6, 417-427.

Nyberg, L., Sandblom, J., Jones, S., Neely, A. S., Petersson, K. M., Ingvar, M., & Backman, L. (2003). Neural Correlates of Training-Related Memory Improvement in Adulthood and Aging. *PNAS,* 100(23), 13728-13733.

Obler, L. K., & Fein, D. (1988). *The exceptional brain: neuropsychology of talent and special abilities.* New York: Guilford Press.

O'Brien, D. (2000). *Learn to remember: practical techniques and exercises to improve your memory.* San Francisco: Chronicle Books.

Ong, W. J. (1982). *Orality and literacy: the technologizing of the world.* London: Methuen.

Osborne, L. (2003, June 22). Savant for a Day. *New York Times.*

Peek, F., & Anderson, S. W. (1996). *The real rain man, Kim Peek.* Salt Lake City, Utah: Harkness Publishing Consultants.

Petroski, H. (1999). *The book on the bookshelf.* New York: Alfred A. Knopf.

Phelps, P. (n.d.). Gender Identification of Chicks Prior to Hatch. *Poultryscience.org e-Digest,* 2(1).

Pinker, S. (1994) *The language instinct: how the mind creates language.* New York: W. Morrow and Co.

Radcliff-Ulmstead, D. (1972). Giulio Camillo's Emblems of Memory. *Yale French Studies,* 47, 47-56.

Ramachandran, V. S., & Hubbard, E. M. (2001). Psychophsyical Investigations into the Neural Basis of Synaesthesia. *Proc. R. Soc. London,* 268, 979-983.

Ramachandran, V. S., & Hubbard, E. M. (2003, May). Hearing Colors, Tasting Shapes. *Scientific American,* 53-59.

Ravennas, P. (1545). *The art of memory, that otherwyse is called the Phenix A boke very behouefull and profytable to all professours of scyences. Grammaryens, rethoryciens dialectyke, legystes, phylosophres [and] theologiens.*

Ravitch, D. (2001). *Left back: a century of battles over school reform.* New York: Simon & Schuster.

Rose, S. P. (1993). *The making of memory: from molecules to mind.* New York: Anchor Books.

Rose, S. P. (2005). *The future of the brain: the promise and perils of tomorrow's neuroscience*. Oxford, England: Oxford University Press.

Ross, P. E. (2006, August). The Expert Mind. *Scientific American,* 65-71.

Rossi, P. (2000). *Logic and the art of memory: the quest for a universal language.* Chicago: University of Chicago Press.

Rowland, I. D. (2008). *Giordano Bruno: philosopher/heretic.* New York: Farrar, Straus and Giroux.

Rubin, D. C. (1995). *Memory in oral traditions: the cognitive psychology of epic, ballads, and counting-out rhymes.* New York: Oxford University Press.

Sacks, O. W. (1995). *An anthropologist on Mars: seven paradoxical tales.* New York: Knopf.

Schacter, D. L. (1996). *Searching for memory: the brain, the mind, and the past.* New York: Basic Books.

Schacter, D. L. (2001). *The seven sins of memory: how the mind forgets and remembers.* Boston: Houghton Mifflin.

Schacter, D. L., & Scarry, E. (2000). *Memory, brain, and belief.* Cambridge, Mass.; London: Harvard University Press.

Shakuntala, D. (1977). *Figuring: the joy of numbers.* New York: Harper & Row.

Shenk, D. (2001). *The forgetting: Alzheimer's, portrait of an epidemic.* New York: Doubleday.

Small, G. W. (2002). *The memory bible: an innovative strategy for keeping your brain young.* New York: Hyperion.

Small, G. W., & Vorgan, G. (2006). *The longevity bible: 8 essential strategies for keeping your mind sharp and your body young.* New York: Hyperion.

Small, J. P. (2005). *Wax tablets of the mind: cognitive studies of memory and literacy in classical antiquity*. London: Routledge.

Smith, S. B. (1983). *The great mental calculators: the psychology, methods, and lives of calculating prodigies, past and present.* New York: Columbia University Press.

Snowdon, D. (2001). *Aging with grace: what the nun study teaches us about leading longer, healthier, and more meaningful lives.* New York: Bantam.

Spence, J. D. (1984). *The memory palace of Matteo Ricci.* New York: Viking Penguin.

Spillich, G. J. (1979). Text Processing of Domain-Related Information for Individuals with High and Low Domain Knowledge. *Journal of Verbal Learning and Verbal Behavior,* 14, 506-522.

Squire, L. R. (1987). *Memory and brain.* New York: Oxford University Press.

Squire, L. R. (1992). *Encyclopedia of learning and memory.* New York: Macmillan.

Squire, L. R., & Kandel, E. R. (1999). *Memory: from mind to molecules.* New York: Scientific American Library.

Standing, L. (1973). Learning 10,000 Pictures. *Quarterly Journal of Experimental Psychology,* 25, 207-222.

Starkes, J. L., & Ericsson, K. A. (2003). *Expert performance in sports: advances in research on sport expertise.* Champaign, IL: Human Kinetics.

Stefanacci, L., Buffalo, E. A., Schmolck, H., & Squire, L. (2000). Profound Amnesia After Damage to the Medial Temporal Lobe: A Neuroanatomical and Neuropsychological Profile of Patient E.P. *Journal of Neuroscience,* 20(18), 7024-7036.

Stratton, G. M. (1917). The Mnemonic Feat of the "Shass Pollak" *Psychological Review,* 24, 244-247.

Stromeyer, C. F., & Psotka, J. (1970). The Detailed Texture of Eidetic Images. *Nature,* 225, 346-349.

Tammet, D. (2007). *Born on a blue day: inside the extraordinary mind of an autistic savant : a memoir.* New York: Free Press.

Tammet, D. (2009). *Embracing the wide sky: a tour across the horizons of the mind.* New York: Free Press.

Tanaka, S., Michimata, C., Kaminaga, T., Honda, M., & Sadato, N. (2002). Superior Digit Memory of Abacus Experts. *NeuroReport,* 13(17), 2187-2191.

Thompson, C. (2006, November). A Head for Detail. *Fast Company,* 73-112.

Thompson, C. P., Cowan, T. M., & Frieman, J. (1993). *Memory search by a memorist.* Hillsdale, N.J.: L. Erlbaum Associates.

Treffert, D. A. (1990). *Extraordinary people: understanding savant syndrome.* New York: Ballantine.

Wagenaar, W. A. (1986). My Memory: A Study of Autobiographical Memory Over Six Years. *Cognitive Psychology,* 18, 225-252.

Walker, J. B. R. (1894) *The comprehensive concordance to the holy scriptures.* Boston: Congregational Sunday-School and Publishing Society.

Walsh, T. A., & Zlatic, T. D. (1981). Mark Twain and the Art of Memory. *American Literature,* 53(2), 214-231.

Wearing, D. (2005). *Forever today: a memoir of love and amnesia.* London: Doubleday.

Wenger, M. J., & Payne, D. G. (1995). On the Acquistion of a Mnemonic Skill: Application of Skilled Memory Theory. *Quarterly Journal of Experimental Psychology,* 1(3), 194-215.

Wilding, J. M., & Valentine, E. R. (1997). *Superior memory.* Hove, East Sussex, UK: Psychology Press.

Wood, H. H. (2007). *Memory: an anthology.* London: Chatto & Windus.

Yates, F. A. (1966). *The art of memory.* Chicago: University of Chicago Press.

INDEX